LEEDS BECKETT UNIVERSITY
Leeds Metropolitan University
17 0219552 5

Business and Commercial Aspects of Engineering

Business and Commercial Aspects of Engineering

JOHN HUNT BSc (Hons), CEng, MIEE

A member of the Hodder Headline Group
LONDON • SYDNEY • AUCKLAND

First published in Great Britain 1997 by
Arnold, a member of the Hodder Headline Group,
338 Euston Road, London NW1 3BH
http:\\www.arnoldpublishers.com

British Library Cataloguing in Publication Data
A catalogue record for this book is available from the British Library

ISBN 0 340 67667 1

Publisher: Dilys Alam
Production Editor: Liz Gooster
Production Controller: Rose James

Typeset in 10/12 pt Palatino by Saxon Graphics Ltd, Derby
Printed and bound in Great Britain by J.W. Arrowsmith Ltd, Bristol

Contents

Preface

This book is written to provide the material necessary for GNVQ Unit 1: Engineering and commercial functions in business (Advanced). The aim is to introduce students to the engineering and commercial functions relevant to them, and demonstrate the interfaces and relationships between them. No prior knowledge of the general scope of engineering in businesses is assumed nor is any prior commercial knowledge necessary.

The text has been structured to conform to the specifications of the three unit elements as set out in the GNVQ guidance document. Thus, students and lecturers should find it easy to relate the text and chapters to the required performance criteria, range and evidence indicators of the unit.

Part 1 relates to Unit Element 1.1: Investigate business functions which involve engineers. This section provides an introduction to the role of engineering in business:

- Chapter 1 explains the economic framework of business, business sectors, types of company and the business cycle, thus providing an awareness of the general economic and commercial background to business.
- In Chapter 2, the common business functions are introduced and the possible flat, hierarchical or matrix organisation structures are shown. Business management structures are explained, together with the component management activities such as supervising, planning and controlling.

Part 2 relates to Unit Element 1.2: Investigate engineering and commercial functions in business. This provides an overview of the general place and role of engineering in business:

- Chapter 3 discusses, in detail, the engineering functions found in business. Emphasis is given to their roles and relationship to each other in the flow of the product from R&D work to after-sales engineering support.
- Chapter 4 covers the commercial functions in detail. Marketing, sales, purchasing and distribution activities are shown to be related to the product flow.
- Chapter 5 outlines financial and management accounting and identifies the major financial decision-making processes.

- Chapter 6 examines the interfaces and information flows between the engineering and commercial functions previously described. The impact of information technology (IT) is also discussed.

Part 3 relates to Unit Element 1.3: Apply techniques of monitoring and controlling costs:

- Chapter 7 explains cost elements, types and behaviour within a business.
- Chapter 8 details techniques of budgeting and the monitoring and control of costs.

Appendices for Discounted Cash Flow values and cost headings are also included.

At the end of each chapter there are some Exercises which can be attempted individually or in small groups. Answers are provided at the back of the book.

Part 1

Unit Element 1.1

Investigate business functions which involve engineers

Overview

Engineering is concerned with the creation and support of products, facilities and systems. Thus, engineering activity is present in many organisations throughout the business community and the economy.

Businesses undertake a wide variety of activities in order to meet the demands of their customers. These activities are normally referred to as functions and include such tasks as designing, making and selling products. Hence the part of the business organisation that does the designing is called the design function, that which does the selling is called the sales function and so on. Engineering activities are carried out by a number of functions such as product design, manufacturing engineering and maintenance. Non-engineering activities such as sales, purchasing and finance are known as commercial functions and the objective of this book is to explain the relationships between these and the engineering functions. It is important that engineers understand these relationships because such knowledge will bring a greater commercial awareness which, in turn, can improve personal effectiveness and assist career advancement. Such knowledge can also assist a move to other functions such as production, sales and purchasing where engineers are often employed.

The engineering disciplines employed (e.g. electrical, mechanical and others) will vary from business to business depending on the type of product and the process technologies involved. However, most companies use similar commercial and financial practices which can be applied to any business, regardless of the engineering and other specialist disciplines involved. We can therefore make valid linkages that are generally applicable between normal engineering functions and the commercial functions of business.

Before we proceed to a detailed examination of the interfaces and relationships between engineering and commercial functions, it is important that students have an appreciation of the general background to the economic framework, types of business, the business cycle and business organisations because they have an important influence on the commercial functions. Part 1 examines these topics, Part 2 looks in detail at the major functions whilst Part 3 deals with financial costing.

1

Types of business

1.1 Economic framework of business

In order to understand the commercial aspects of business, it is necessary to appreciate the nature of the economic framework in which business operates. Figure 1.1 shows the basic economic and business transaction cycle.

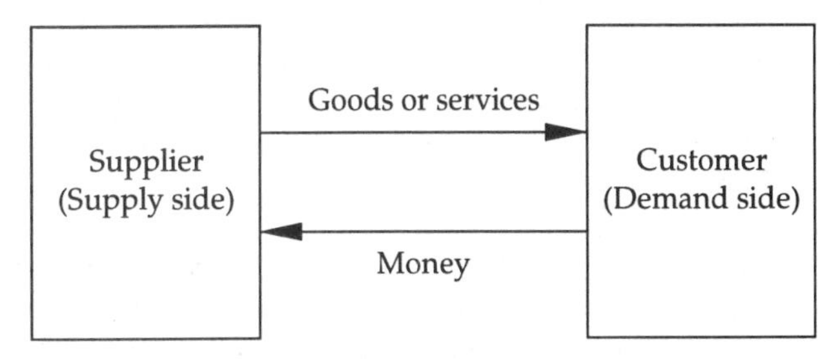

Figure 1.1 The basic economic and business transaction cycle

The concept is very simple but its influence on all business activity is often overlooked by engineers and other non-commercial specialists. The basic cycle illustrated is that goods and services are supplied in return for an agreed sum of money. The transaction can be as simple as buying something for cash over a retail counter or as complex as building a power station to a lengthy contract of agreement. The essential principle of an agreed exchange of value is the same. In fact, the total economic activity of a society is only the daily sum of millions of such transactions between suppliers and customers. We shall see later that a business is essentially a self-sustaining economy in which the functions, departments and individuals are all suppliers and customers of each other.

Business sectors of the economy

The economy is split into a number of sectors that reflect the the economic cycle. The sectors, and the industries and businesses within them, are classified as being either primary, secondary or tertiary sectors.

Primary sector

The primary sector (i.e. the first level) provides the raw materials or basic foods for our society. It includes areas such as mining, agriculture, fishing and forestry that extract, collect or grow the basic materials or foods and sell them to secondary businesses or tertiary businesses.

Secondary sector

The secondary sector (i.e. second level) converts the basic materials or foods into the manufactured products that we buy, e.g. cars, electrical goods, packaged foods, clothing and educational materials. Manufacturing and construction companies make up the bulk of secondary sector businesses.

Tertiary sector

The tertiary sector (i.e. the third level) is otherwise known as the service sector and includes the retail trade which sells the products of the secondary sector to private consumers. Other services include transport, insurance, banking, hotels, communications and holidays. These provide services or facilities that are hired or used for a period but do not normally supply an article to be owned by the consumer. Repair and maintenance services are an exception, where in addition to hiring expertise to effect a repair, some replacement materials may have to be purchased.

The second half of the twentieth century has seen the rapid expansion of tertiary (service) businesses in the economy as general prosperity has grown and our requirements as consumers have moved from basic goods to more diverse, complex and specialist services.

Care must be taken when analysing businesses to see which sector they fall in because some seem to carry out activities in two sectors. For example, some large food and clothing retailers claim to manufacture items of the stock that they sell and these are known as 'own brand items'. In fact, many of these items are made by established manufacturers in the secondary sector under contract to the tertiary retailer whose own name appears on the label. The key issue is, which is the predominant activity of the business?

The interaction of the sectors in the economy

Figure 1.2 shows the theoretical direction of the flow of materials from the primary to the tertiary sectors, but this is a much simplified concept. If we examine the actual flows from the sectors to their potential customers as illustrated in figure 1.3, we can see how interdependent all sections of the economy are. All three sectors supply each other and the private consumer in a closed circle of business and economic activity. This is why a financial event like a change in the Bank of England base interest rate can have such a wide impact nationally through its 'knock-on effect' within the economic circle.

The employment of engineers in the three business sectors

Engineers are employed in all three business sectors. In the primary industries they are to be found in mining, metals production and energy produc-

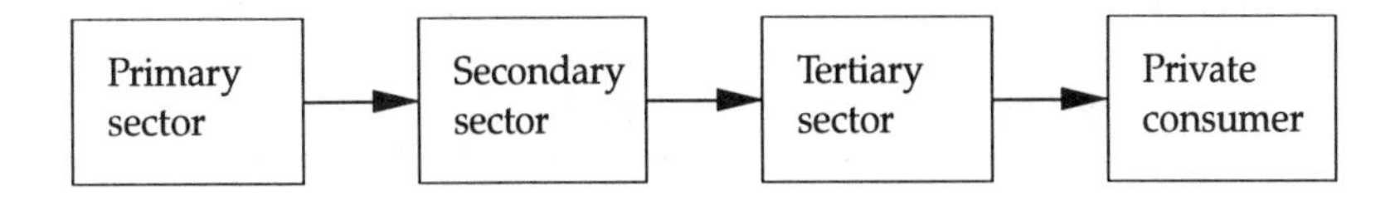

Figure 1.2 Business sectors

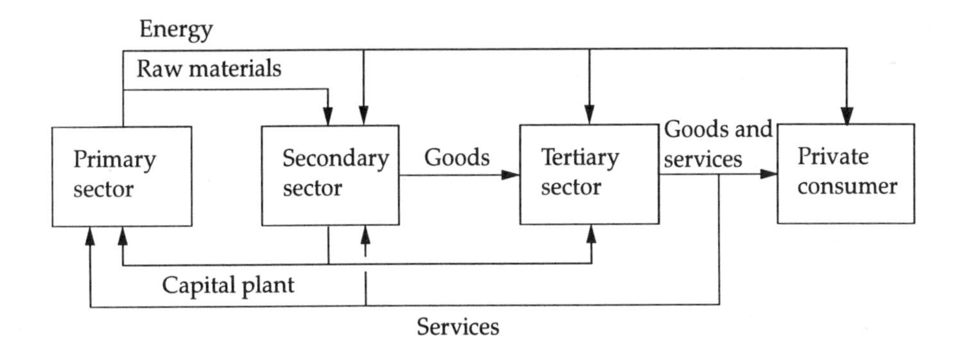

Figure 1.3 Economic flows between business sectors

tion and distribution. In the secondary sector, manufacturing industry employs every engineering discipline whilst the construction industry sees civil, mechanical and electrical engineering used in design, installation and maintenance. Engineers are also employed in tertiary industries such as transportation (railways and airlines), communications (telephone and IT systems), retail outlets (electrical goods and cars) and repair businesses as well as IT work and site maintenance for commercial enterprises such as banks and insurance companies.

International trade

World trade has become far more competitive in recent years due to the widespread use of modern production technologies, computers, satellite communications, bulk transportation by sea and the greater use of air cargo. These facilities have opened up world-wide markets to all trading nations, especially those that are concentrated in the three main trading blocks:

- European Union (formerly the Common Market);
- North America (the USA, Canada and Mexico);
- Far East: Japan and the four 'tiger economies' of South Korea, Singapore, Hong Kong and Taiwan.

The three most powerful trading nations in the world are Japan, the USA and Germany whilst Britain was seventeenth in the league table based on Gross National Product (GNP) of £679 billion in 1994.

It is likely that engineers, through their liaison with commercial departments of their company, will be involved in international trading by applying technical standards to varying foreign market requirements and national import/export regulations. As soon as a business starts importing supplies or exporting products and services, the marketing, sales, design, production, purchasing and distribution functions in particular have extra considerations to deal with.

1.2 Types of business financial structure

The major types of business financial structures in the UK are:

1. The private sector (owned by individuals)
   - public limited companies (plc)
   - private limited companies (Ltd)
   - partnerships
   - sole traders
2. Public sector (owned by the government)
   - public utilities
   - government departments

The size of all types of business except the sole trader can vary considerably but most large companies are public limited companies (plcs) and most of the small ones (under 100 employees) are private limited companies. The difference will be explained below. Public utilities such as the railways, electricity companies and water boards were owned by the government but have recently been 'privatised' i.e. returned to the private sector through the issue of shares to the stock market.

The private sector

Public limited companies (plc)

Public limited companies are owned by shareholders who have bought shares of the business on national Stock Exchanges. The shareholders may be private individuals but pension funds and other businesses such as insurance companies are large investors in these markets. The key Stock Exchange companies are large investors in the stock markets. The key Stock Exchanges world-wide are located in London, New York, Tokyo, Frankfurt, Paris and Hong Kong. Shares are bought and sold in these markets and anyone can buy these shares. This is the feature that makes the businesses public companies.

The shareholders are the nominal owners of the company but there are too many of them for them to act as an effective management. Consequently, public companies have a Board of Directors elected by the shareholders to run the company on their behalf. Some directors may have a significant shareholding and are in that sense 'part owners' whilst other directors have no shares and are essentially managers with director status because of the importance of the business functions that they control. Boards of public companies will also include non-executive directors who do not participate in the day-to-day management of the company but act as independent stewards on behalf of external bodies such as the financial institutions, trade associations and government departments who are important contacts of the company.

A public company must have a minimum of two directors, but usually there will be a larger number of directors, each responsible for one of the main operating functions of the company such as sales, finance, production and engineering. They will control teams of managers within those functions. A Board of Directors is headed by a Chairman but the day-to-day running of the company is normally controlled by a Managing Director. These two posts are usually separate in public companies. Figure 1.4 shows the Board of Directors of a large public company with several separate businesses to be managed. The board will include some directors representing the group or corporate finances and long-term planning, the MDs of the operating businesses and a number of non-executive directors. The company secretary, who also attends, may or may not be a director.

Private limited companies (Ltd)

Private companies have the sale of their shares restricted by agreement between the existing shareholders. Many small companies are private, particularly if they are family-owned. Many public companies were originally private but changed to public status in order to attract additional capital from the general public via the issue of their shares on the Stock Exchange. In other regards, private companies are organised much as public companies but there will be more occasions where a founder or dominant family shareholder is the Chairman and/or Managing Director. Figure 1.5 shows a board structure which is applicable to a private company, smaller public company or the operating divisions of large public companies. In these cases, the board comprises the directors who manage a core function of the business and who report to the MD.

The designation 'Ltd' is used to distinguish private limited companies from public limited companies (plc). 'Ltd' is an abbreviation of the word 'limited' which was originally applied to both private and public limited companies. Private companies need have only one director but must have a company secretary with legal responsibilities for the financial administration of the company. One advantage of a private company is that it does not have to reveal so much information in its annual accounts as plcs do.

Public and private companies are created by the process of incorporation. Incorporation in the UK is regulated by a series of Companies Acts passed by Parliament and companies have to be registered with the Registrar of Companies. The type of business and powers of the company have to be

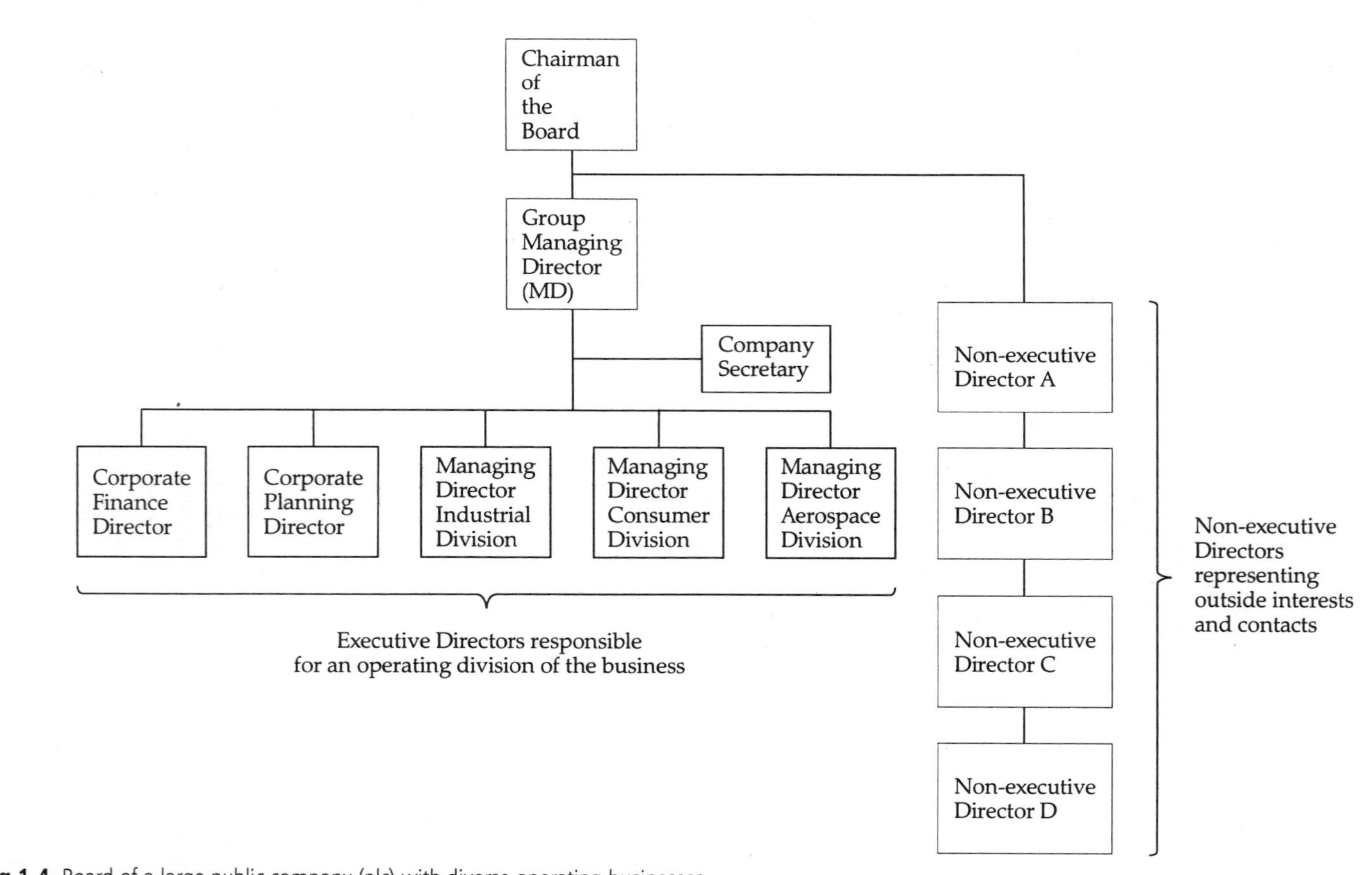

Fig 1.4 Board of a large public company (plc) with diverse operating businesses
The Chairman of the Board can be executive or non-executive. The MD is the chief executive officer (CEO) who runs the day-to-day business.

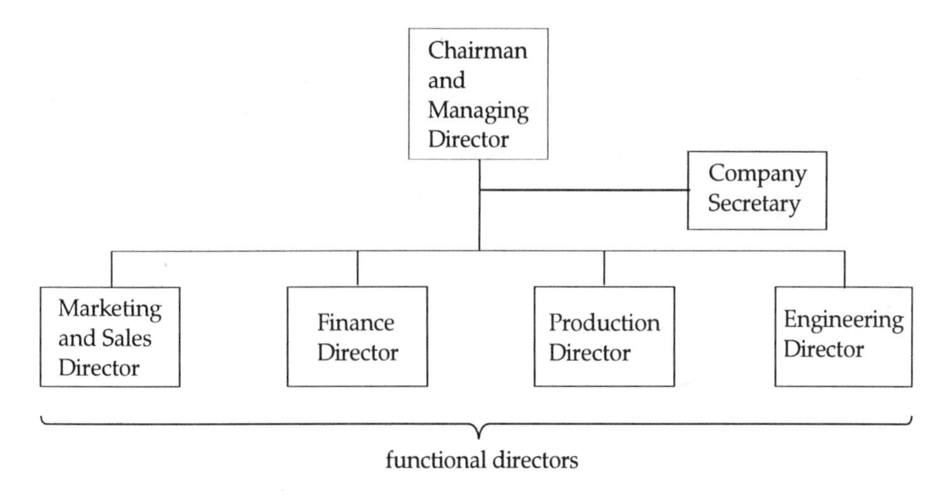

Figure 1.5 Board of a medium-sized business or one with a single operating organisation

specified in a Memorandum of Association whilst the powers of directors and internal controls are specified in Articles of Association.

Partnerships

Partnerships consist of two or more individuals who agree to work together as one business. In principle, the partners agree to share the assets and liabilities of the business equally and if one partner defaults, the other(s) can be held responsible for the liabilities. This type of organisation is common in professional engineering practices such as design and consultancy services. However, many such businesses are now limited companies in order for the members to have the personal protection that limited liability gives. Some professions such as solicitors and accountants are not allowed in English law to become limited companies. Partnerships also do not have to reveal as much information in their accounts as plcs do.

Sole traders

Sole traders are individuals working on their own account. Many will have the tax status of self-employed but a significant number register as private limited companies for financial and marketing reasons, and legally become employees of the company that they own. Many engineers working individually as a professional engineer or skilled crafts registered as either self-employed or as a private limited company. Sole traders do not have to disclose their accounts publicly.

1.3 The business cycle

In order to understand the following chapters on business organisations and their commercial and financial work it is important that engineering students have a knowledge of the basic business cycle. We will illustrate this by describing a business where the cycle for the design, production and repeat sale of one product type might be:

- Design and/or specify the product in detail.
- Obtain orders for the product.
- Purchase the materials needed.
- Make the product.
- Test the product.
- Deliver the product.
- Invoice the customer for payment.
- Receive payment.
- Pay for the materials and any services used.
- Pay wages.

Fig 1.6 shows the sequence as a cycle where payment for the first deliveries pays for purchases and work on further orders. The sequence can vary depending on the type of product and the commercial environment. For example, detailed design or production may not start until AFTER the customer has ordered the product, This is known as 'design or make to customer's order'. On the financial side, materials may have to be paid for upon delivery or within one month depending on the creditworthiness of the purchaser and the selling policy of the supplier. The proprietor may, therefore, have to pay suppliers for materials and services BEFORE receiving payment for the finished product.

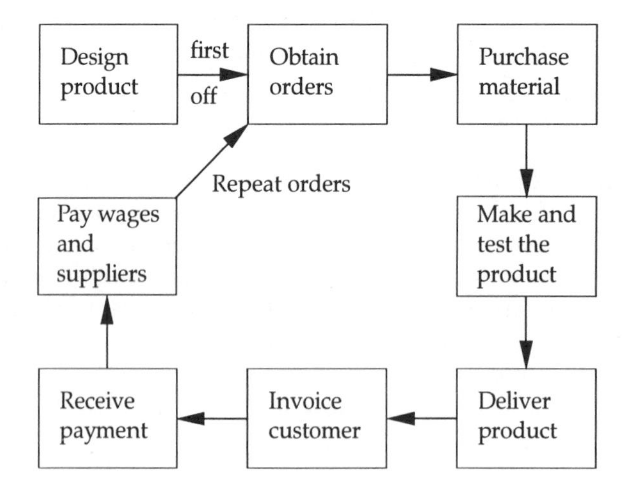

Figure 1.6 Business cycle for a product with repeat order

Business life is rarely as simple as the single cycle shown in Figure 1.6. The first likely complication is that more than one sales order may be processed at the same time, but they are at different stages in the business cycle because orders are received and deliveries promised at different times. With more than one current order to deal with, the proprietor or manager has four options:

1. Work on the first order only and not start the second until the first is completed.
2. Stop work on the first order and start and complete the second order.
3. Work on both orders alternately for a period.
4. Employ more people so that more orders can be worked on at the same time.

The first option is the most appropriate where jobs are short in duration or have to be done in a continuous sequence for technical reasons. The second option may please the second customer but annoy the first. The third option is only practical if the cycle for each order has waiting times built in that allow alternate working, or when the delivery targets for both orders allow sufficient time to work alternately. The fourth option is viable only if the additional employment is temporary to cover only the second order or further sales orders can be gained that will warrant the permanent employment of additional staff.

1.4 Exercise 1

1. Which business sector is your company in and why is this so?

2. Give three examples of businesses in each sector.

3. Which type of business financial structure is the most common and why is this so?

4. What are the most likely financial structures for a new small company and what the options?

5. Explain the concept of limited liability.

6. Detail the basic business cycle for a design and manufacturing company.

7. Detail three factors that are likely to affect the sequence of basic cycle in a business.

8. Examine the basic business cycle shown in Figure 1.6 and divide the functions into groups of similar work. You should identify a minimum of three groups.

2

Business functions and organisation

2.1 Business functions

Question 8 in Exercise 1, Chapter 1 asked you to look at the tasks of the business cycle and see if you could put any of them into groups of similar work. We can identify that some of the tasks that were concerned with the creation and detailed design of the product, others with its manufacture and still others with the financial transactions needed. This grouping of like-tasks creates basic functions that are found in most companies. The functions present in any business will depend on the type of business and product being sold, but they would be drawn from:

- Marketing
- Research and Development (R&D)
- Product Design
- Purchasing
- Manufacturing
- Production Control
- Inventory Control
- Stores
- Manufacturing Engineering
- Quality Assurance
- Sales
- Distribution
- Finance
- Personnel

Many of the functions found in businesses are closely related and are often constituent departments of a larger function. Table 2.1 shows some typical functional groupings within a large design and manufacturing company. Smaller businesses will undoubtedly combine some of the constituent departments with the core functions depending on their relevance and the manpower affordable.

The list is not a blueprint for the ideal business but a comprehensive example of the sort of functional relationships that exist in many companies. Each business will have a unique functional structure best suited to its circumstances. Most design and manufacturing businesses will have the recognisable core functions and it is a matter of size and emphasis as to how necessary it is to split these into constituent functions. The sole trader will carry out all the necessary functions personally, whereas in a large company many of the staff will be specialists employed in core and constituent

Table 2.1 Functional groupings

Core function	Constituent function
Marketing	Market research New product planning Product promotion
Product design	Research and development Design office Drawing office Applications engineering sales support
Manufacturing	Production departments Production planning and control Inventory control and stores Manufacturing engineering Plant engineering
Quality assurance	Quality Control Quality Engineering and laboratory Inspection departments
Purchasing	Production materials Other goods
Sales	Sales areas Distribution warehouses
Finance	Financial accounting Management accounting
Personnel	Personnel administration Training

functions. The major point to understand is that most of the functions will need to be carried out in some form in any such business whether it is large or small.

Other types of company from different business sectors will employ some of the functions but have a different emphasis on them. For example, retail shops will omit design and manufacture, but substitute a greater emphasis on the selection, wholesale purchasing, stockholding and display of goods to sell.

Bearing the business cycle and the basic functions in mind, we can now explore the different kinds of business organisation, using a secondary sector company that designs and manufactures an engineering product as our example.

2.2 Types of organisation

There are five types of organisation to which businesses conform:

1. Flat
2. Hierachical
3. Matrix
4. Centralised
5. Decentralised

These can be illustrated by diagrams called organisation charts which we shall use in the following sections.

Flat organisations

Some organisations have very few levels of staff or function and are therefore regarded as 'flat'. The simplest example is of a business owned by one proprietor with a small number of staff reporting to him or her. In this case there are only two levels of organisation – the proprietor and the staff, all of whom report directly to the proprietor and are of equal status.

Even in a flat organisation there are two ways of arranging the tasks of the business cycle to decide the question of 'who does what'. The options are:

1. All the staff take an order each and do everything required in the business cycle to complete that order.
2. Divide the work into similar tasks and organise the staff to do only one or a limited number of tasks, but to do it for all the orders.

The second option is by far the most common because it has the advantage of people becoming expert in one area of activity. Figure 2.1 shows the flat organisation defined by people and Figure 2.2 the same organisation defined by function.

Hierarchical organisations

As businesses grow larger to accommodate more orders, so the numbers of people in each function will increase. It is logical that as this happens people will specialise in only one or a few of the tasks undertaken by the function. For example, a company might start with one salesperson who will respond to all sales enquiries but as these grow in number an additional salesperson is recruited. One might then specialise in UK orders and the other in export orders. Alternatively, if the product sales needs specialist technical knowledge, one salesperson may deal with mechanical products and the other with electrical products.

The principle of specialisation can be applied to all functions. As a company grows in size, the amount of knowledge, data and communications required increases to a point where it is more efficient and safer to have people specialise in one area of knowledge and activity. Consequently, as the

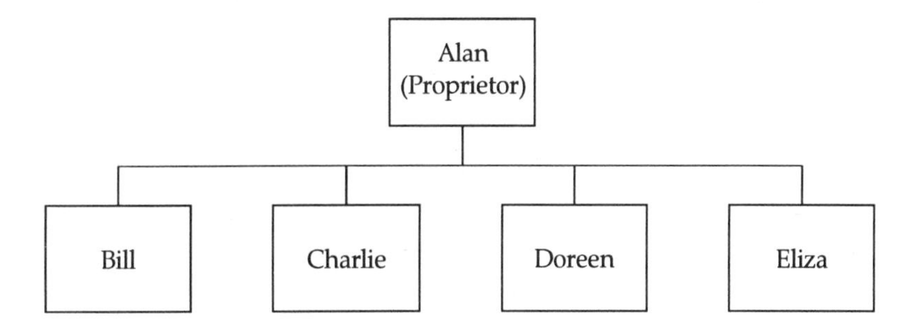

Figure 2.1 Flat organisation by staff

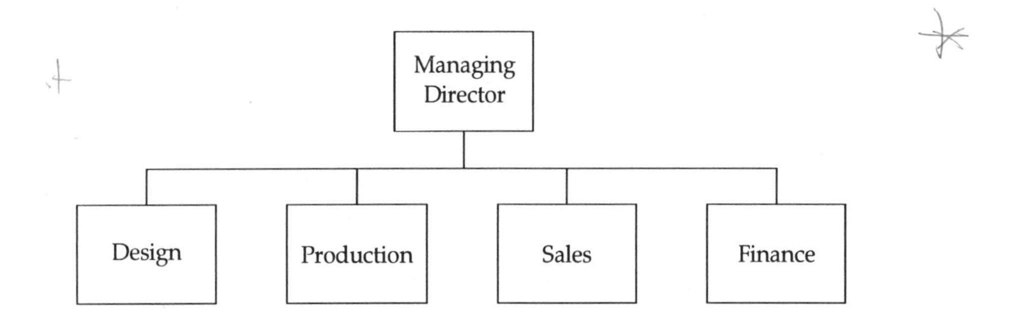

Figure 2.2 Flat organisation by function

size of the business increases so do the number of specialist departments or sections within the main functions. Figure 2.3 shows a hierarchy where specialist departments have developed within the main functions. The departments themselves would be sub-divided into sections as the need arises.

The advantage of the hierarchical organisation is the concentration of competence in a formal organised management structure. In recent years, however, problems have arisen with the increasing complexity of business and the need to respond more quickly to changes in very competitive markets. The problem is illustrated by Figure 2.4 which shows that, although policy and decisions flow vertically up and down the organisation structure, the actual progress of an order from receipt to the finished product is horizontal across the structure. Although much of the communication of data between staff at all levels is horizontally across the organisation, that is it flows with the business and product cycle, the need for formal decision making up and down the vertical structure can cause time delays in response and misunderstandings of the detailed situation.

Matrix organisations

Matrix organisations are modifications of the hierarchical structure designed to address the above problem by creating a more cohesive

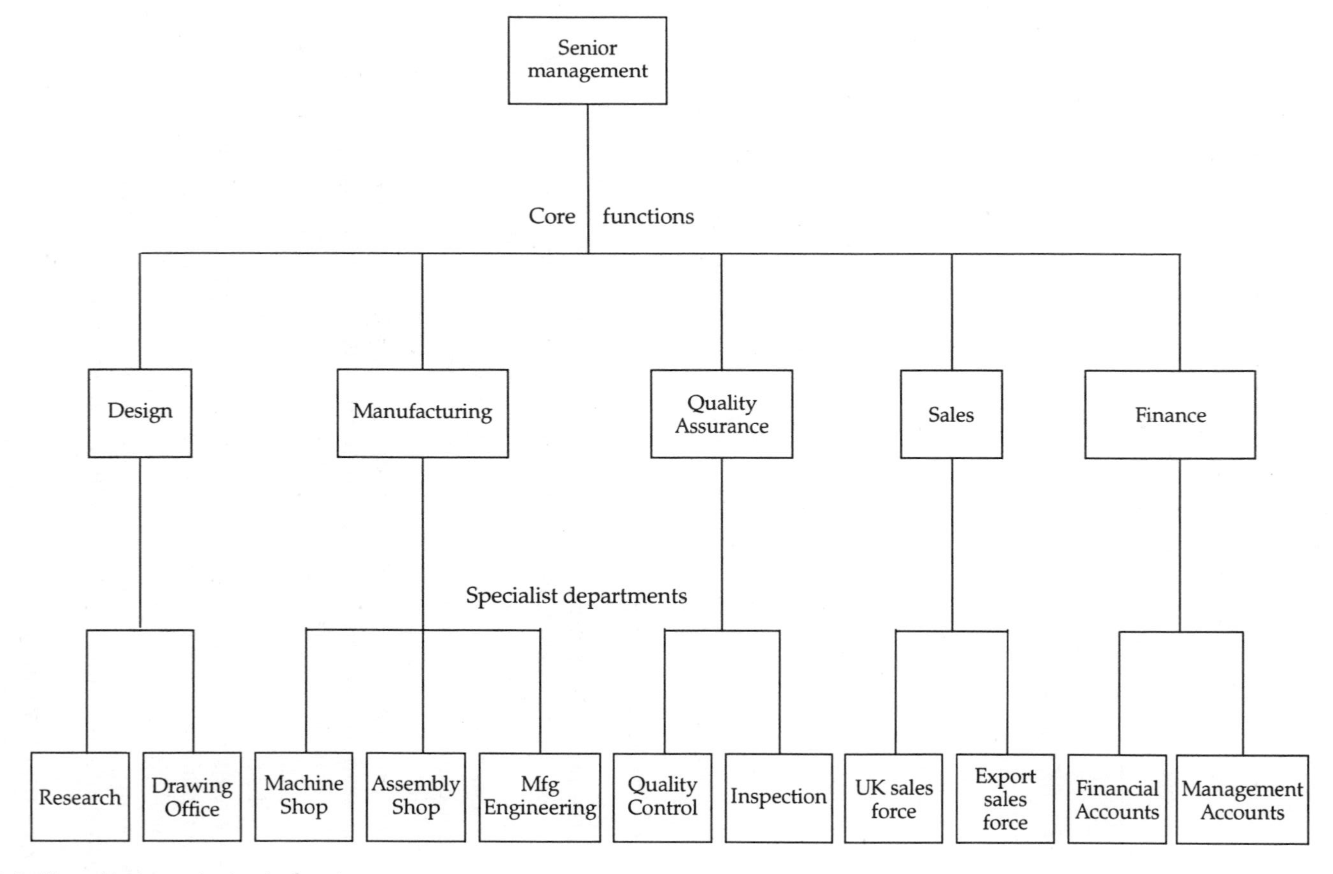

Figure 2.3 Hierarchical organisation by function

business-wide input into an order, product line or one-off project. It does this by establishing temporary teams or permanent cells of staff to concentrate on a particular product or project throughout its entire business cycle or a substantial part of it. Figure 2.5 shows a matrix organisation where staff who are nominally employed by a particular function are seconded to a project team or profit centre.

This type of matrix has been used for large one-off capital or operational projects for many years but is now being applied to give a permanent change to the organisation structure. It requires both an organisational and geographical change. In a pure hierarchical system accountants, salesmen, turners, assemblers and engineers work in separate locations and the jobs flow to and from them. In a permanent operational matrix the specialists are drawn from their functions and work together as a team in one location. For example, instead of having an assembly shop and a machine shop producing separately and being supported by engineers located in far-off offices, all the facilities and people needed for the production and support of a product or project are located together or as near as possible. There may be resource and geographical limitations placed upon this principle but the basic objective is to get as close as possible to a cohesive multi-disciplined team located in one site. In other words, to create 'a factory within a factory'.

One of the major benefits of matrix management has been to increase the personal identification and commitment of staff to the end product and whole business ethic. In hierarchical organisations people are isolated from the other operations in the business cycle but in matrix organisations people can see a direct relationship between what they do and the whole business cycle. The matrix also reflects the wider industrial realisation that when dealing with people the assumed economies of scale gained by big organisations have ignored the advantages of the increased motivation that people get from being a valued part of a focused team and not just 'a small fish in a large pond'. Normally staff still remain in a function for employment purposes because it allows a commonality of standards of recruitment and practice in each specialism and gives management the flexibility to allocate staff from the function pool of labour as operational changes require it.

Another outcome of introducing matrix organisation is that it devolves more management responsibility to the staff of the teams, units or cells. In doing so, this improves personal motivation and inter-disciplinary communication.

Centralised organisations

Centralised organisations are structured so that all the functions of the company report directly to one central head office or administrative centre for senior decision-making. Figure 2.6 shows such an organisation. In theory, the advantages of centralisation have been held to be:

- Maximum management control from the centre (especially financial control).

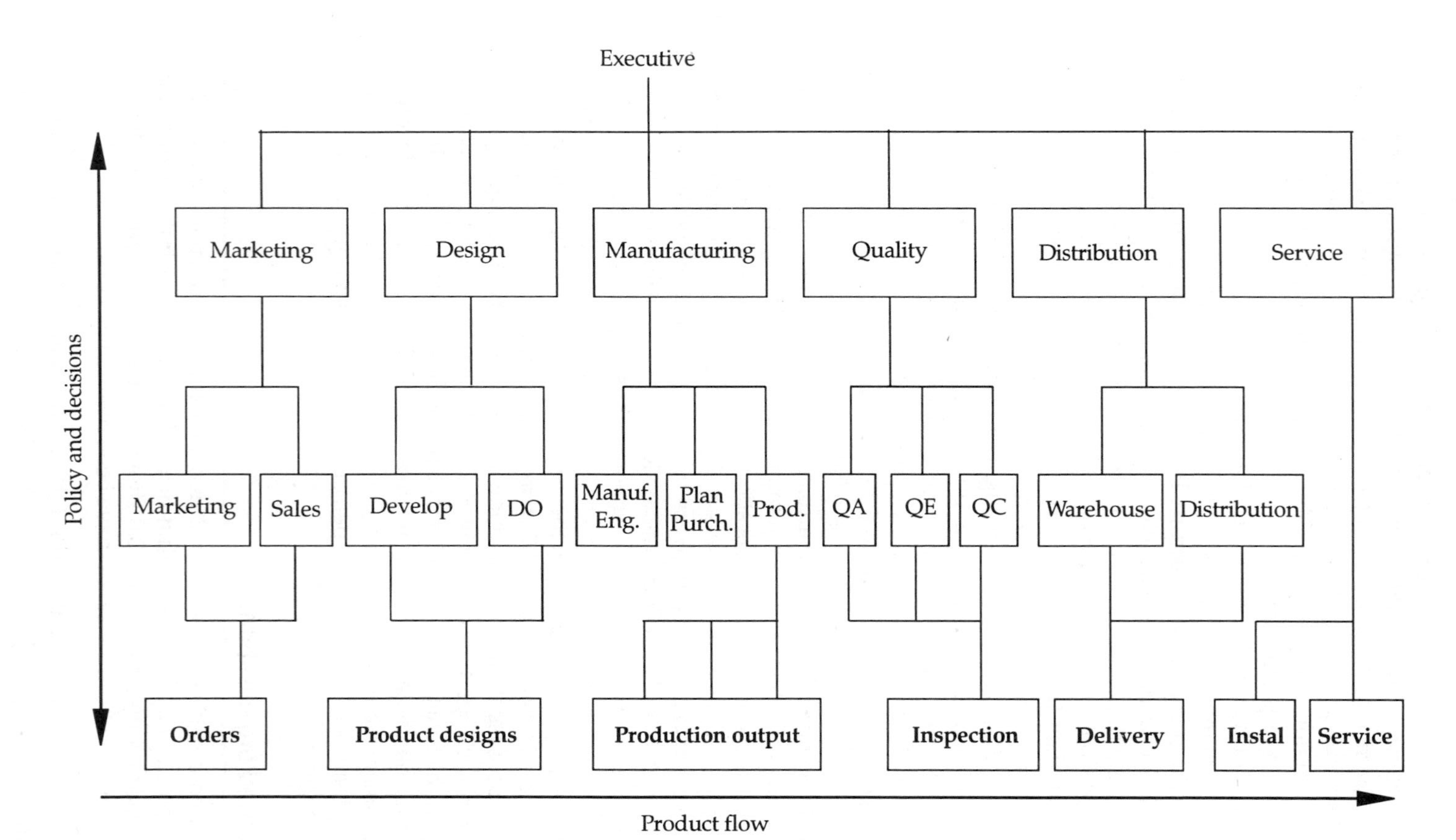

Figure 2.4 The flow of products and decisions in hierarchical organisations
DO = Design Office, QA = Quality Assurance, QE = Quality Engineering, QC = Quality Control

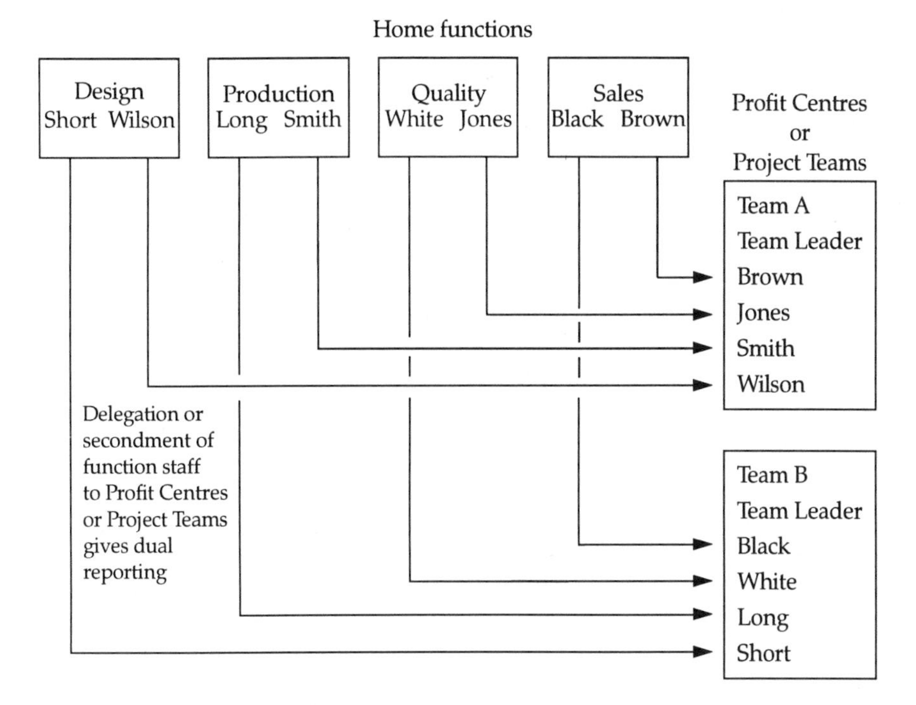

Figure 2.5 Matrix organisation

- Better decision-making through the wider corporate viewpoint of the top decision-makers.
- Greater operational commonality in functions across the organisation.
- A greater corporate sense of identity and purpose.
- Economics of scale from having fewer separate functions for the same total level of corporate activity. In theory, creating a flatter organisation.
- A useful control technique for growing or parent companies to absorb and re-direct the activities of other companies that have been taken over.

The disadvantages of centralisation are as follows:

- Although in theory centralisation needs fewer levels of organisation, what has happened in many businesses is that it has created extra levels of administration to co-ordinate the separate operational units into a whole. For example, a number of national sales territories or a number of factories need to report to an additional level of corporate executive.
- Decision-making can be slow because it has to go up and down the management chain through managers who are overloaded with such a wide remit.
- Decisions are poorer because they are made by managers remote from the local conditions who therefore cannot grasp the finer points of the arguments.

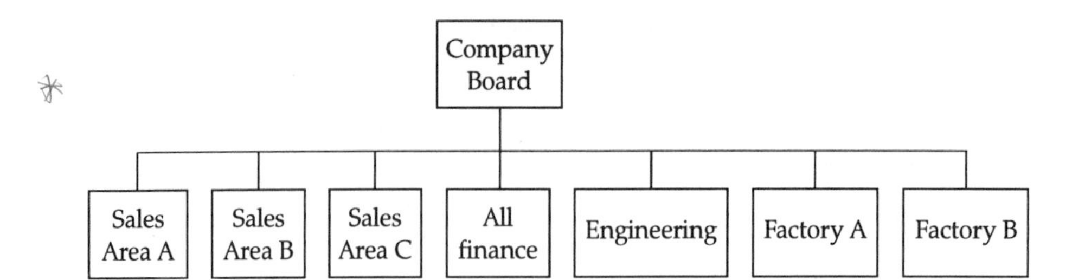

Figure 2.6 Centralised organisation

- This in turn leads to a lower level of personal identification and motivation of local staff right across the corporation because they have no control over their situation.

A great move to centralisation of organisations occurred in the high economic growth period of the 1960s and 1970s when companies expanded and conglomerates combining businesses from widely differing industries grew up. One of the developments that aided the centralisation trend was the installation of central mainframe computers and software programmes that aided the central control of corporate finance, sales and inventory.

Decentralisation

Figure 2.7 shows a decentralised organisation. In the more competitive economic situation of the 1980s and 1990s the organisational trend has been for many conglomerates to divest themselves of peripheral businesses and to concentrate on their original core activities. The recession has also forced businesses to decentralise.

The advantages of decentralisation are:

- Decentralisation gives more authority for decision-making at the point where the decision is applied. This gives better decision-making because of the greater local knowledge of the business and issues involved.
- Decision-making is quicker because of the fewer levels of management involved.
- Administrative costs can be reduced by the elimination of co-ordinating levels of functional or area management. Only the operational levels of management are left.
- Staff identify better with local senior management and are more motivated because they have greater influence on their situation.

A list of the disadvantages of decentralisation will be very similar to the above list of the advantages of centralisation on p.17–19.

In business and industry, however, there has been a definite trend towards the breakdown of large centralised structures into more independently managed operating units. The conclusion drawn from this must be

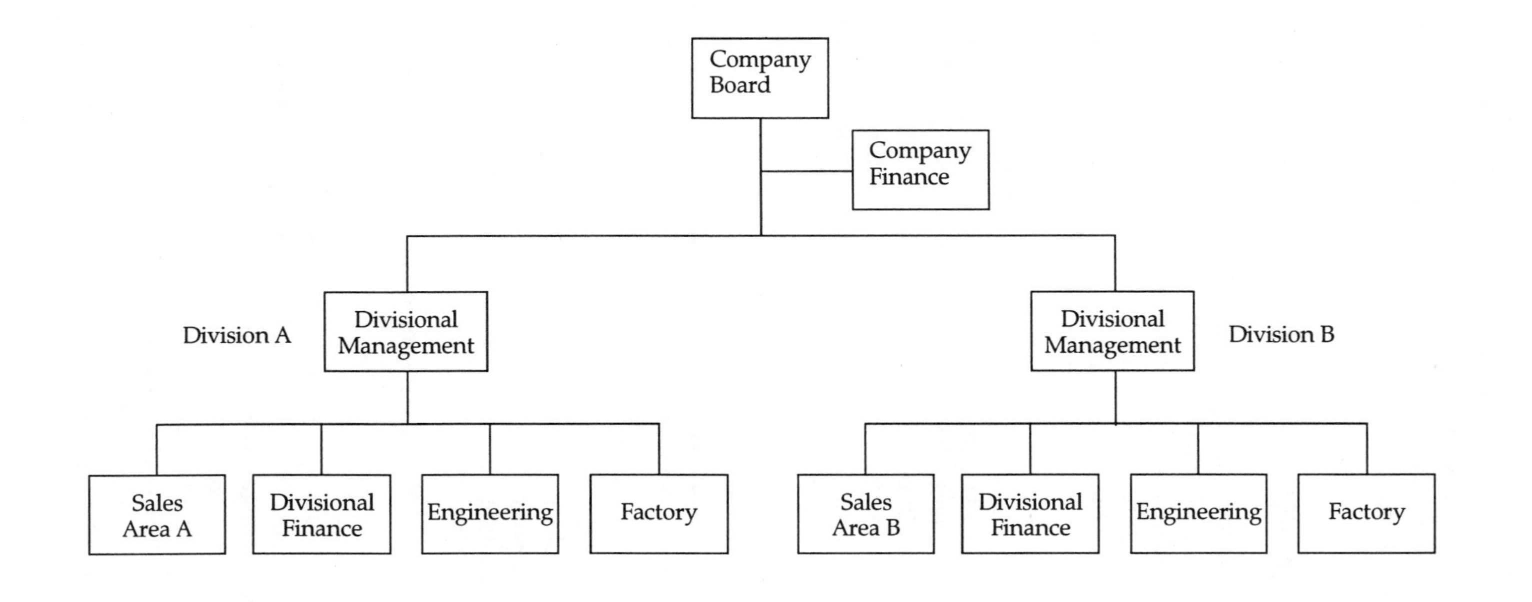

Figure 2.7 Decentralised organisation

that under the more competitive environment of the 1980s and 1990s practical experience and economic necessity have guided businesses to favour decentralisation of structure and the accompanying delegation of authority and responsibility to the lowest possible levels in the organisation in order to gain the advantages listed above.

2.3 Business management

A business has to be managed efficiently if it is to survive and succeed. Thus managers need to be positioned in various parts of the company to control the activities in that area. We have examined typical business functions and organisation structures in the previous sections and it is logical for businesses to use these as the basis for the placement of managers and supervisors. Consequently, an organisation structure is also a representation of the management structure and the seniority of managers is confirmed by their place in the structure, i.e. it reflects the relative size and scope of their responsibilities. The factors examined in this section apply to all managers and management whether they are in engineering or commercial functions. This is an important point because it provides a common working methodology and language for managers and their staff that adds understanding when working with people from other functions. This section looks first at general management activities and then at typical management roles.

Management activities

During the conduct of business a number of common activities are carried out by management in all functions of the company. These are:

- Supervising
- Managing
- Directing
- Deploying
- Organising
- Planning
- Controlling
- Forecasting
- Facilitating
- Co-ordinating

Some of the terms used have very similar meanings and the activities can be grouped together.

Supervising encompasses the overall control of a working section, group or project, but is sometimes understood to refer particularly to the guidance and control of people. Use of the term supervising is not restricted to describing the activities of supervisors, but is a general term that equally describes the activities of managers when managing and directors when directing. Supervising is usually based on the application of some level of official authority over the supervised. For example, a supervisor will have the authority to issue instructions to staff and expect them to be implemented. Supervising will involve carrying out all of the following activities at some time.

Deploying involves the general allocation of resources, but again, in particular to the assignment of people to tasks. It is one of the main responsibilities of supervision. Deployment can be short-term in allocating labour for only a few minutes or hours or long-term in assigning a person to a project for, say, a two-year contract for a customer.

Organising involves the orderly arrangement of activities and resources. This can include the deployment of people and the securing of supplies and equipment to come together in a programme of activity to an agreed timetable or plan.

We will now consider *planning* and *controlling* which when linked together form a fundamental tool of good business management. Planning is the activity of deciding an objective and how it is to be achieved, normally within a target time period. Planners need to have detailed knowledge of the possible methods of achieving the targets set and the extent of the current resources available to do so. The key element of planning activity is the comparison of objectives to the resources available in a given time period. For example, a production plan will specify the quantities, methods and delivery of a range of items from a production department, bearing in mind the productive capacity of the department. Planning is carried out at all levels in the organisation by a cascading process up or down the organisation and management structure. Top-down planning takes the company plan and then successively breaks it down into function, department and section plans. Bottom-up planning builds plans up by taking the lowest levels of requirement (the section plan) and successively combining them into departmental and functional plans.

Controlling is the activity of overseeing an event or programme to ensure that the targets are being met or seem likely to be met upon completion. The key elements in controlling are:

- Measuring the actual results.
- Comparing the measurement with the target.
- The feedback of information to the planning activity so that corrective action can be taken if there is a variance between the measured result and the targets. Figure 2.8 shows this diagrammatically.

The cycle of planning–measuring–feedback–controlling to meet the original target in changing circumstances is a fundamental activity in business. It often involves considerable resources. For example, Production Planning and Control departments are set up in factories to carry out the cycle as routinely and smoothly as possible for the manufacturing function. The cycle is undertaken by all business functions in some form of planning and control system, and the financial budget control systems of businesses is a prime example. It is also a cycle that is carried out continuously and often automatically by all of us in our daily lives. The task of crossing the road is just one example of the importance of planning and controlling through information feedback.

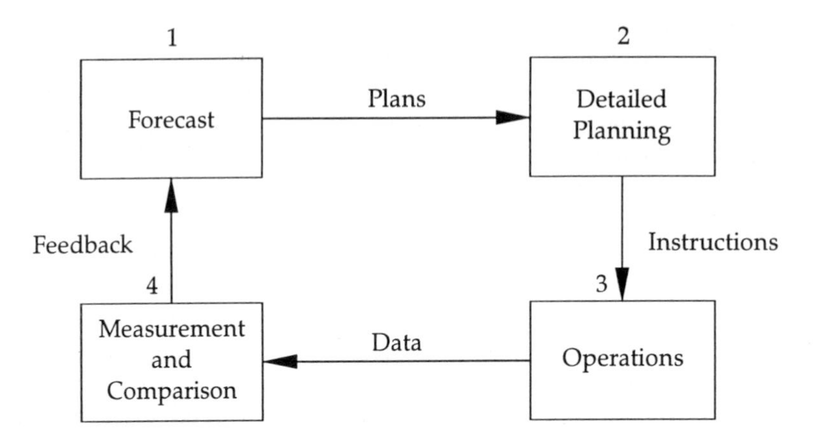

Figure 2.8 Planning and control loop

Forecasting is an activity that predicts future requirements based on assumptions about possible influencing factors. Forecasts are used by planning staff to prepare detailed programmes of action. For example, a sales forecast produced by the Sales function will be used by Production Planning to set the production schedules of the factory. In another example a maintenance manager will have to forecast the probable usage of machine spare parts to ensure that they are on site when a machine breaks down. Forecasting uses various statistical methods that are examined in section 8.3.

Facilitating is an organisational activity that arranges conditions so that persons involved in tasks or activities are given the maximum opportunity to succeed through their own efforts. It involves supportive action by managers and others in charge of time and resources to put these at the disposal of people. For example, a team of employees looking at how they might improve their operation will be given time, a room, presentation materials and, most importantly, access to information they might not see in the normal course of their work.

Co-ordinating is an element of organisation and supervision. It seeks to ensure that all working efforts are best arranged to meet an objective. Time is usually an important consideration in co-ordinating because in any venture success depends on the right things happening at the right time. For example, if a new machine is to be installed without interrupting surrounding production for too long (if at all), it will be important to co-ordinate the availability of the installation engineers with the delivery of the machine.

Features of management

So far in this chapter we have examined the usual functions of a business, typical organisation structures of the functions and the activities commonly carried out by all supervisors, managers and directors. This last section examines the structural features of management and gives profiles of management roles at three levels in a typical organisation.

Authority and delegation

Management responsibility is largely determined by the arrangements for the exercise of authority and its delegation. Authority is the power to make decisions and direct the work of others, and is normally conferred by the position of a manager in the organisation structure. The maximum number of persons that a supervisor can be expected to control effectively on a full-time basis is known as the span of control and this is said to be seven. Delegation is the passing down of authority to a subordinate who then has the powers of direction of the manager. For example, when a manager is on holiday, his or her general responsibilities may be delegated to a deputy but there would be limitations applied. Typically the deputy would have general powers for day-to-day operational decision-making but would not be authorised to hire or dismiss personnel.

Line and staff management

There are two types of managerial position: line management and staff management. Line management is the logical progression of authority from each level of management to the next lower one. At each level the manager controls all the functions of the lower levels of management. Figure 2.9 gives an example of line management structure.

Staff management has more limited powers. A staff manager is usually responsible for a particular aspect or activity but does not have the full control of a line manager. For example, a manufacturing engineer working permanently within a production area will have staff responsibility and authority on technical decisions and can instruct production workers and setters in those matters, but has no authority over them for their discipline and general supervision. This remains the responsibility of the Production Manager who is the line manager in the sense used here.

Financial and Personnel department managers often operate in a staff management role because they provide particular expertise to the whole organisation and therefore to every manager. Their input is advisory but carries the authority of their expertise and the need to establish common company-wide practices in these areas. It can be seen from this that staff management is a matter of maintaining a careful balance between the relevant managers. A line manager cannot afford to disregard the expert advice or decision inputs of the staff manager but at the same time the staff manager must not be seen to be undermining the overall authority of the line manager. Figure 2.10 illustrates staff management responsibility.

Three management profiles in engineering

We will now illustrate the major aspects of management by giving a profile of three levels of engineering management in a manufacturing company: junior, middle and senior managers.

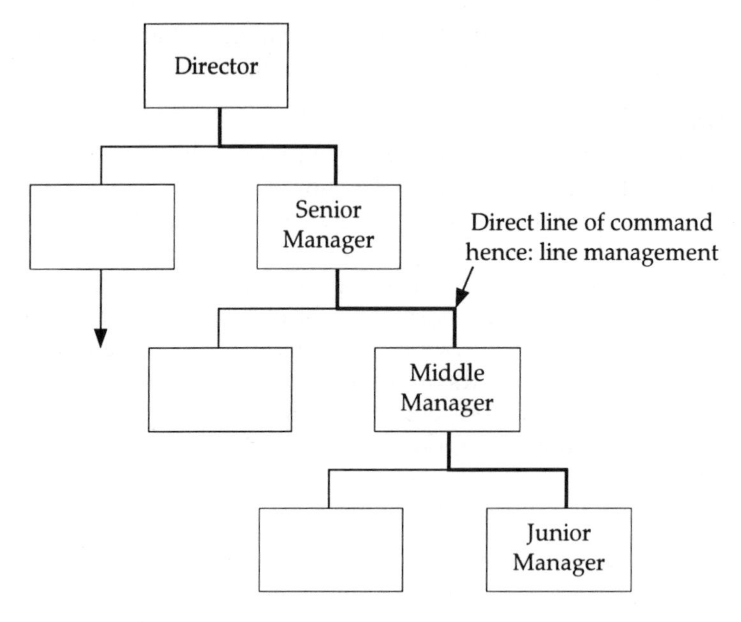

Figure 2.9 Line management structure

Profile of a supervisor or junior management role

- Scope: responsible for a section or small department. Directly supervised staff will be engineers and technicians. The primary distinction between sections is likely to be the technical discipline involved. e.g. mechanical fitting or electrical fitting; alternatively, product A assembly or product B assembly.
- Workload: planning and control will be on a daily and weekly basis. Involvement with detailed work allocation and technical issues that arise will form a large part of the role.
- Practical knowledge and hands-on experience of the specialism are still a major requirement at this level but personnel and management skills are needed to supervise others. Financial control, personnel recruitment and discipline will normally be the responsibility of a superior, or require the approval of a superior, i.e. middle manager.

Profile of a middle management role

- Scope: responsible for an engineering department comprising a number of sections. Directly supervised staff will be section supervisors or junior managers and perhaps some departmental clerical support. Either technical discipline (e.g. mechanical engineering) or engineering function (e.g. Product Design Department) can be the distinguishing factor between departments.

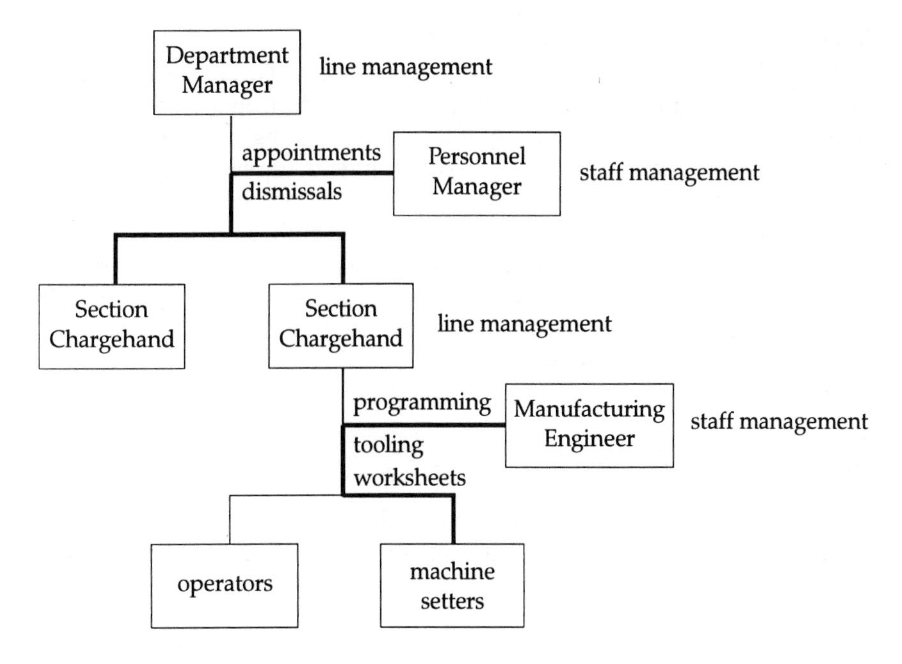

Figure 2.10 Staff management structure

- Workload: Planning and control are on a weekly and monthly basis. The time spent on planning, financial controls and personnel matters increases.
- Liaison with other departments and senior management for the co-ordination of plans, objectives and responses to problems can take up a significant proportion of the time available. Involvement with daily technical details is reduced, although significant technical problems or new developments will need the full attention of the manager.

Profile of a senior management or director role

- Scope: is responsible for an engineering function with number of departments. Directly controlled staff will be middle managers plus possibly specialist and clerical support. Planning and control are on a monthly, quarterly and annual basis and are part of the company or unit overall plan.
- Workload: the senior manager will be responsible for the overall performance of the engineering function and the success of major engineering projects. His main focus will be the proposal, approval, planning and overall control of major engineering programmes, new product development, departmental budgets and capital expenditures.
- Financial control and budgeting for the total engineering function will be significant. Meetings with other senior managers, subordinate managers and personnel from other sites, customers and suppliers will occupy a majority of his or her time.

Senior engineering managers and the CEO

Meetings with the Chief Executive Officer (CEO) will occur to discuss major engineering programmes, customer problems and budgets. These will normally be infrequent in many companies apart from regular scheduled meetings of all senior managers. This is because, having agreed on a product and engineering policy, the Chief Executive's daily preoccupations are with finance, sales, marketing and production output rather than the underlying enabling technologies. Only in the consideration of new product development programmes or where engineering performance is causing a serious problem to the other operational areas or customers is the CEO likely to be involved with detailed engineering issues.

2.4 Exercise 2

1. Consider the core functions outlined in the text and draw up lists of functions which might comprise the constituent sub-functions of the following businesses.

   (a) A company with six employees in the retail sales of spares for motor cars.
   (b) A company with 25 employees, sub-contracting machine parts for a number of manufacturers.
   (c) A company of 100 staff designing and manufacturing an electronically controlled fluid pump.
   (d) A company of 500 staff producing the same product.
   Draw an organisation chart in each case.

2. What are the two major functions apart from Sales in a design and manufacturing company? Name three of the primary tasks for each of the functions.

3. Draw a table showing the inclusion or not of the following functions in the three types of business given below: R & D; Product design; Purchasing; Manufacturing; Manufacturing engineering; Sales; After-sales support; and accounts.
   (a) Sub-contract machining.
   (b) Design and manufacture of washing machines.
   (c) Design of printed circuit boards.

4. How many levels of management and supervision are generally appropriate for the following numbers of staff?: 10; 50; 500; 1000.

5. Discuss the advantages of communications and decision-taking between the management organisation structure of a design and manufacturing business, on one hand, and the flow of the product through the organisation on the other.

6. Define the following types of organisation structure.
   (a) Flat.
   (b) Hierarchical.
   (c) Matrix.

7. Which type of organisation is best for:
   (a) A business of 20 people producing one product.
   (b) A business of 500 people producing a wide range of products.

8. What is meant by centralised and decentralised organisations?

9. Discuss the main features of:
   (a) Supervising.
   (b) Deploying.

10. Discuss the relationship of controlling activities to planning activities.

11. What is the difference between facilitating and co-ordinating?

12. Explain the difference between line management and staff management. Give examples in the form of a simple organisation structure.

Part 2

Unit Element 1.2

Investigate engineering and commercial functions in business

Overview

In business, engineering activities operate within two types of commercial environment.

1. Where the product sold is an engineering product such as cars, televisions or aircraft. In these cases many of the company functions will be concerned directly with engineering work and the commercial functions will be dealing with issues related to the engineering of the product.
2. Where the product sold is not an engineering product. Examples are food and drink manufacture, hotels and airlines, in which case engineering activity is limited to providing and maintaining site services or production facilities of some kind. The commercial functions will have little connection with engineering-related issues.

The distinction can be illustrated by the difference of emphasis between the manufacturer and user of an engineering product, in this example an aircraft manufacturer and a commercial airline. The aircraft manufacturer is involved in a large amount of engineering design and engineering production to exacting technical standards, in order to manufacture an aeroplane with operational reliability, fuel efficiency and an extremely high level of in-flight safety. This involves a predominance of engineering or engineering-related activities across the organisation. On the other hand, the airline is primarily concerned in selling seats on its aircraft and managing the tremendous logistical problems of organising planes, aircrews, passengers, food and fuel to a flying schedule. In this case the engineering maintenance of the aircraft to high standards is very important for the safe and regular operation of the aircraft, but even so this engineering function forms only a part of a much wider highly commercial and logistical transportation business.

In the following chapters we will focus on businesses producing an engineered product because these provide the most comprehensive examples of the relationship between engineering and commercial functions.

3

Engineering functions

3.1 Introduction

The application of engineering skills within businesses is usually very wide, with a number of functions employing engineers from a variety of disciplines such as mechanical, electrical and manufacturing engineering. For example, electrical engineers could be working in product design, manufacturing engineering, plant maintenance, sales, purchasing and other functions. Conversely, departments concerned with product design, manufacturing engineering or quality assurance may include chemical, electrical, manufacturing, and mechanical engineers in proportions appropriate to the product and its manufacture.

Increasingly, in matrix organisations multi-disciplined teams are brought together and the engineering knowledge and technical responsibilities of individuals are more integrated with colleagues than in the past. This is particularly so where different disciplines are combined into project teams or an engineering department to meet the specific needs of the company and its engineering functions.

3.2 Core engineering functions

The wide variety of engineering activities found in business can be classified as being within one of four core engineering functions, which are identified by the technology areas involved:

1. Product design and sales support.
2. Manufacturing process design and support.
3. Site facilities design and support.
4. Quality Assurance (QA) and support.

Figure 3.1 shows a possible organisation chart for a large company utilising all the core functions and their major constituent functions. The mix and contribution of these in a business will depend primarily upon the degree of

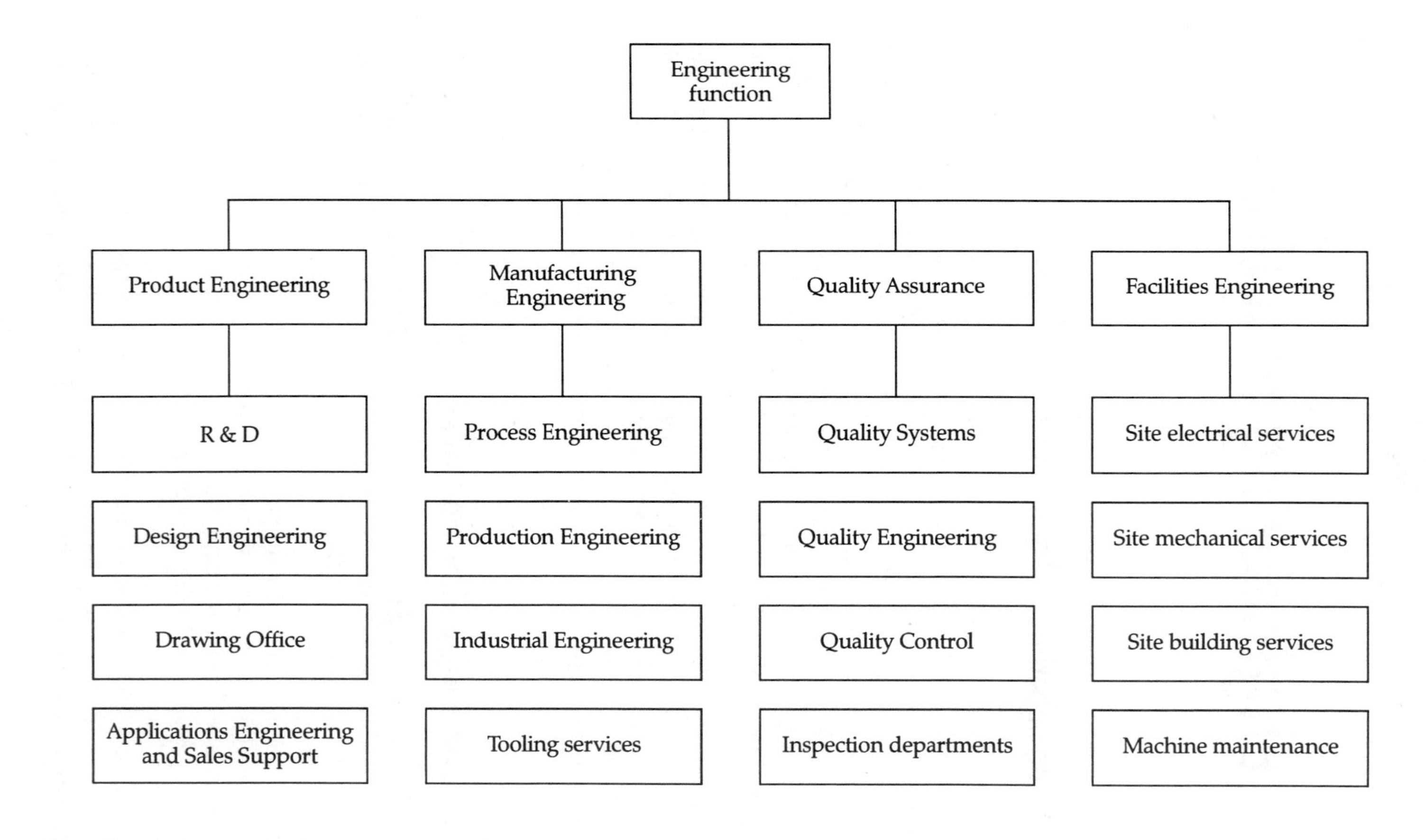

Figure 3.1 Core Engineering functions

design and manufacturing activity involved. Typical combinations of engineering functions that are found in businesses are:

- *A large design and manufacturing company* (e.g. domestic appliances) will have all four functions, each with its own staff and manager who may or may not report to the same senior executive. QA reporting is a special case and we discuss this more fully in Chapter 6.
- *A smaller design and manufacturing company* (e.g. industrial equipment) may combine some of the management roles in engineering because of the lower number of engineers involved. Whether this is desirable from the technical standpoint depends on the criticality of the engineering and its possible influence on product quality and safety. Typically a combination (at management level) of manufacturing engineering and facilities engineering may report to a Production Director whilst product design and QA report to a Technical or Engineering Director.
- *A design consultancy* (civil engineering, tooling or PCBs) housed in an office block will obviously have a design function but also QA for the control of drawings, revisions and design standards. CAE and CAD systems support will be the company's responsibility but engineering support for the site facilities, for example, power and heating will usually be provided by the building landlord.
- *A sub-contract-supplier* (e.g. machined components) will have manufacturing engineering to programme CNC machines and provide the tooling, plus quality assurance (nowadays the absence of QA with BS5750 approval means no orders for the company). Facilities engineering may be a department in a large company or 'the maintenance man' in a small one. Design engineering will not normally be present but the production of very complex and expensive parts requires constant liaison between the company's manufacturing engineers and the customer's part designers.
- *A design and installation company* (e.g. a heating and ventilating supplier) comes into the facilities engineering function. It will have a design function that specifies a heating and ventilating system incorporating specialist capital equipment made by other companies. Electrical and mechanical building services and heating and ventilating suppliers fall into this category. They usually have sales engineers for initial contact and the submission of proposals prior to detailed design, a design function and an installation and test engineering function to install and commission the systems.
- *A sales and service company* (e.g. vehicle or industrial equipment distributor) may have product sales engineers to sell the product and service engineers on site or visiting clients' premises if warranty or service contracts are applicable. It may also have some facilities engineering staff to support its distribution warehouses and showrooms.

The examples above illustrate the wide range of possible contributions by engineering functions in business. We will now examine the functions in greater detail and determine the structures and contributions made within them.

3.3 Engineering functions and titles used

The titles of engineering departments carrying out the same activities vary from organisation to organisation. We shall now examine the common titles in use within the four core functions.

Product design and sales support

Common activities and titles used include:

- Research
- Development
- Research and Development (R&D)
- Product Design Office
- Drawing Office (DO)
- Project Engineering
- Applications Engineering
- Sales Engineering
- Field Service
- Customer Service

The last four entries in the second column of the above list are sales-related engineering support activities. They are grouped in the design function for examination in this text because their common aim with the design function is to supply the customer with the required product and to ensure its satisfactory life in service. Their common language is product knowledge (as opposed to manufacturing/facilities engineering knowledge) plus responsibility to the customers for the product performance over, specifically, the warranty period and more generally the normal life span of the product. Figure 3.2 shows a possible organisation chart for a large company using these functions.

The core product technology of any company comprises all the knowledge and activities relating to the research, development, design, performance and in-life servicing of the product. The knowledge is usually housed in functions or departments of similar title although their form and organisation will vary greatly from company to company, depending upon the needs of the individual business.

The constituent functions are listed in the order in which they would contribute to the complete cycle of taking a product from an idea in R&D to being serviced in a customer's plant.

Research and Development (R&D)

Research examines fields of interest or possible applications in the future. These may or may not be related to existing products and technologies used by the company. Research results should point the way for further development work. Development work will normally concentrate on probable future products that may be entirely new items or significant improvements of existing designs.

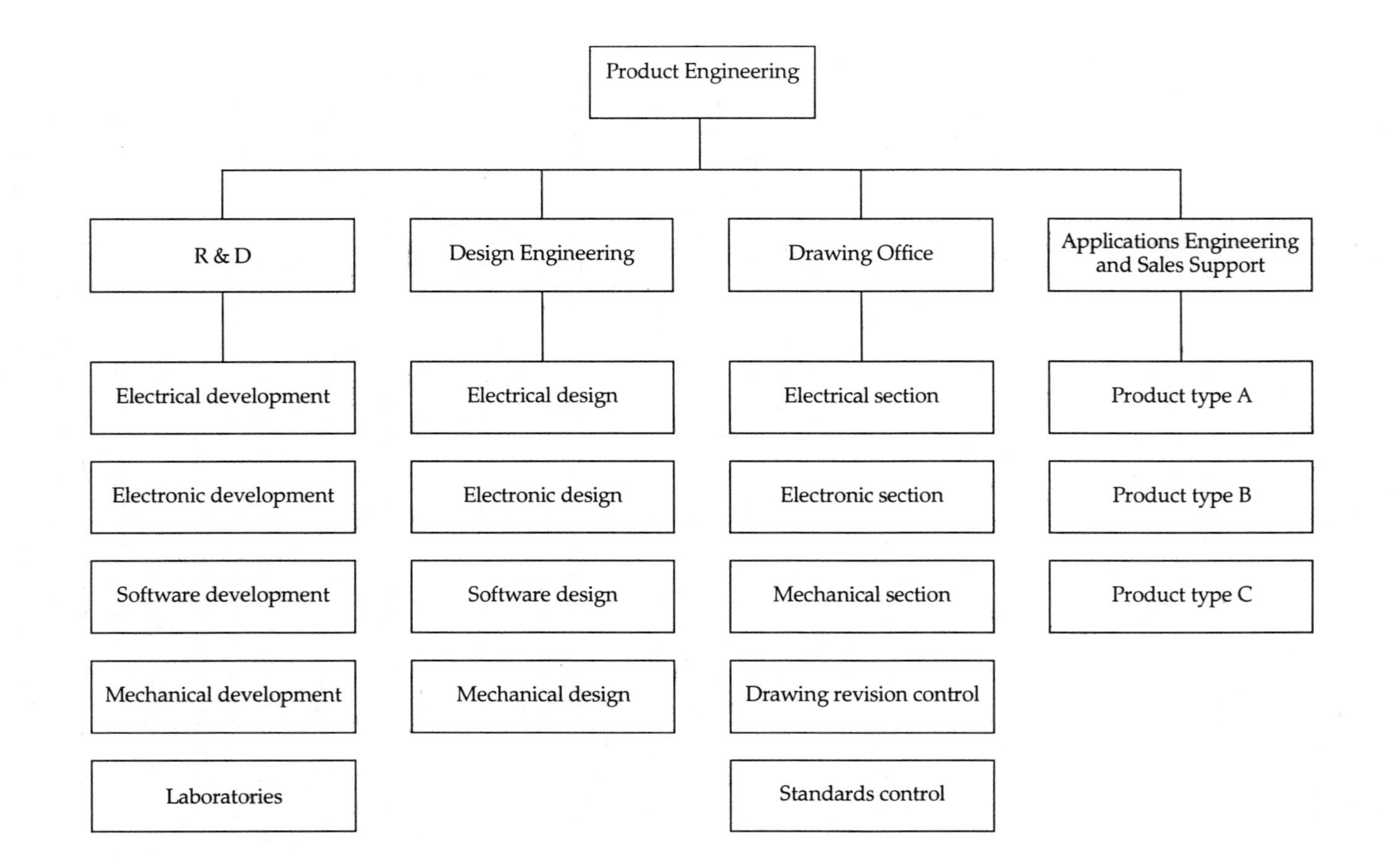

Figure 3.2 Product Engineering organisation chart

Experimental calculations work will lead to the building of prototypes which will be tested for performance. The results will be used to refine the broad specifications before work commences on the detailed product design.

Product Design Office

The Product Design Office produces the final design (usually after considering several options) from design calculations of the product for sale. Top-level design layout, assembly drawings and performance specifications will be produced for further detailing in the drawing office.

Drawing Office (DO)

The Drawing Office can be integral with the design office but there is often a need to have a separate organisation set up for the efficient production of many detailed drawings, material specifications and performance parameters. These will be created in a cascade from top-level assemblies through sub-assemblies down to individual part details and specifications. They will be drawn up for both the manufacture and quality of the product and its performance in use for the customer.

Many companies now use computer-aided engineering (CAE) for design work and computer-aided drawing (CAD) systems for some design and all drawing work. These have software written for the particular type of engineering design task. For example, there are specialist software suites for structural engineering, general mechanical engineering, heating and ventilating, printed circuit board (pcb) and many other applications.

Project Engineering

A large order or project will entail a significant amount of design and development work plus possibly on-site installation and commissioning. Civil engineering and large public works are mainly of this nature and techniques of project management have been developed to manage and co-ordinate the multi-disciplined teams involved. A Project Manager will often co-ordinate a team of Project Engineers with varying expertise who have been brought together to satisfy the requirements of the order. This is an example of matrix management which is now employed in all levels of business where the success of any development or change relies on successful teamwork across the formal organisational structure. Within manufacturing industry the most significant projects are new product developments which involve the whole of the organisation as the project moves from R&D work to the distribution and sale of the product.

Applications and Sales Support

This section provides support to customers and the sales force in selecting the correct model or considering design variants for special applications as well as providing solutions to problems found in service. This function is normally found in businesses supplying specialist capital equipment to other businesses rather than domestic equipment for the consumer markets.

The support will normally be given in conjunction with sales staff in their specification work prior to a sale and with customer service staff in their post-sales support troubleshooting.

Customer Service

This is usually an organisation that provides after-sales assistance and service to customers at their premises. It will be manned by engineers trained to service and repair products on-site and will also supply replacements and spares. It is often known as field service and is usually part of the sales or distribution function. It is included here because of its strong product technology basis.

There should be close contact with the sales department as both are primarily concerned with satisfying the customer and maintaining a continuing contact with him or her. Technical problems that cannot be solved by the service engineers would be referred back to the product/applications engineers within the product design function.

Manufacturing process design and support

Common activities and titles used within this core function include:

- Manufacturing Engineering
- Process Engineering
- Production Engineering
- Industrial Engineering
- Methods
- Estimating
- Jig and Tool Department
- Toolroom
- Toolstores

Figure 3.3 shows the organisation of the manufacturing engineering function of a medium-sized engineering company. A factory will certainly include this core function and some of its common activities along with Facilities Engineering and Quality Assurance. The factory site may or may not house the design function.

The terms manufacturing, production, methods, process, plant and facilities engineering are the most widely duplicated in their application to engineering functions that support the production of the product. For example, a manufacturing engineer in one company would be called a process or methods engineer in another. However in the manufacture of discrete products it is easy to identify the division between engineering for product design (product engineering), the means of production (manufacturing, production or methods engineering) and general factory services (facilities, factory or plant engineering).

Manufacturing Engineering

This is now used as the name of the engineering discipline formerly known as Production Engineering. It is concerned with providing the most efficient

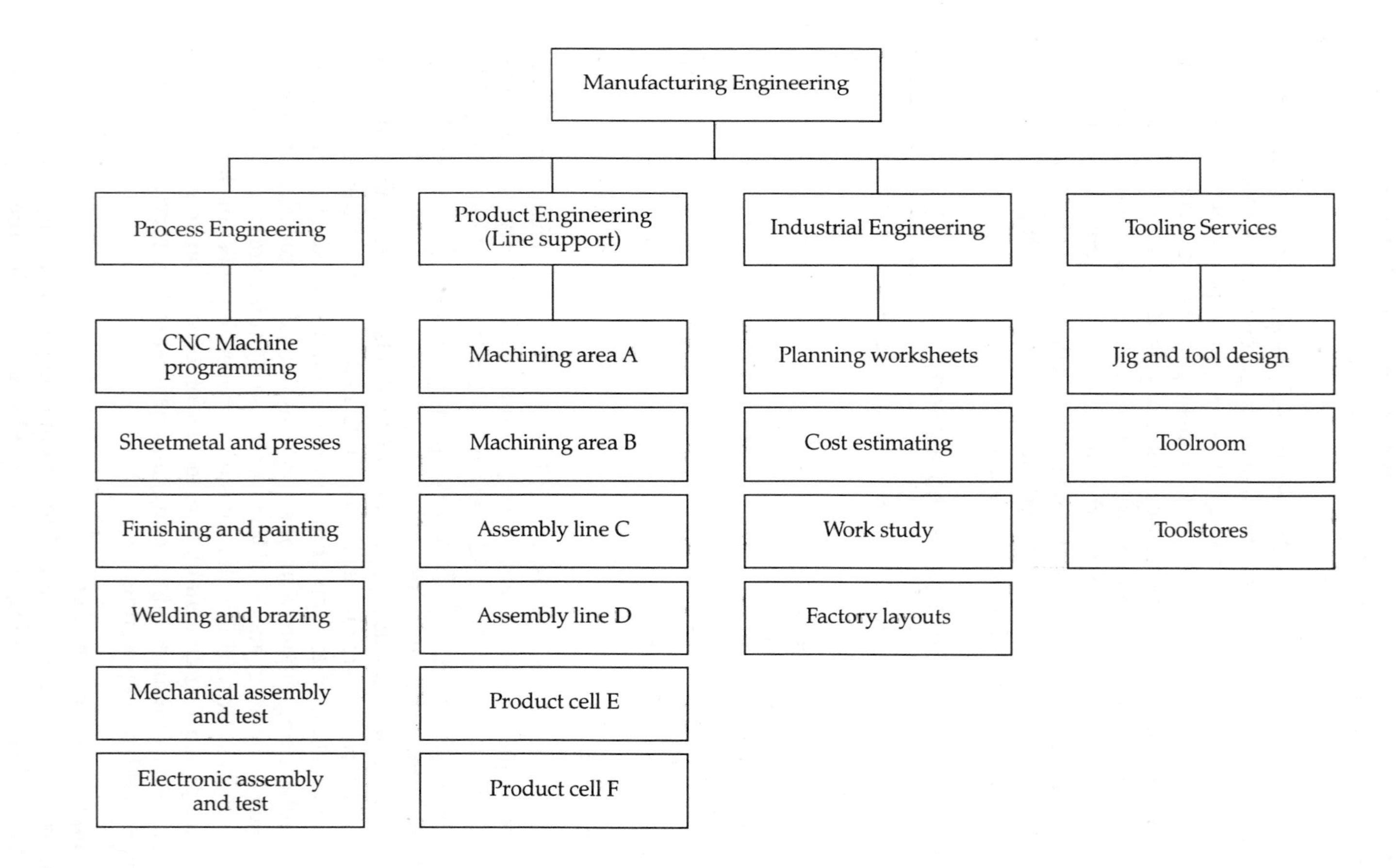

Figure 3.3 Manufacturing Engineering organisation chart

methods and equipment for producing the product to stated quality levels. Machines, tooling and labour methods are analysed, designed and introduced to maximise production efficiency. Manufacturing engineers decide the best method of production relevant to volumes required, quality levels, equipment available, labour requirements, safety requirements, investment policy and the time and costs of alternative solutions. CAM (computer-aided manufacture) systems are used for Computer numerical control (CNC) programme writing and tooling issues in many companies. Production planning and control, shop floor data capture and inventory control systems are also part of the disciplines concern because manufacturing engineers are often involved in or wholly employed in these areas. The manufacturing engineering function should be found in all manufacturing companies where production efficiency, product cost and quality are important.

Process Engineering

Process engineers are specialists in a particular type of process, for example, testing of electronic products, CNC turning or welding. Figure 3.3 shows some of the more common areas of process specialism in engineering companies.

A particular case is in the process industries which are involved in the manufacture of bulk petrochemicals, plastics, drinks and food. Here the main production processes are the application of heat, refrigeration, pressure, vacuum, distillation and mixing of raw materials, in continuous flows through process vessels and connecting piping. These technologies usually involve large-scale structural and mechanical engineering works such as oil refineries and chemical plants. The engineering disciplines involved are predominantly mechanical, electrical and chemical engineering, not manufacturing engineering as seen in factories producing discrete products. Thus, the technologies are more aligned with the facilities engineering described later.

Production Engineering

As mentioned previously, the term has been used in the widest sense to describe all of what now comes under the title of manufacturing engineering. In this text it is used to describe the direct support of production operations by an engineer. In many cases the engineer is resident in the production area and is the first point of reference for any technical problems encountered in the area. The engineer (often called a Line Engineer) would contact the other engineering functions when necessary. The advantage of this is that it leaves the production supervisor (who may not be an engineer in any case) free to concentrate on the supervision and control of labour, materials, production schedules and output.

Industrial and Methods Engineering

This is usually involved in ergonomics and organisation of workplace and has developed from the earlier applications of work study disciplines. Other functions often included in this area are the activities of process planning, cost estimating and workplace and factory floor layouts.

Work study techniques are used to measure accurately the working practices and environment of a process, particularly where it involves labour working independently of a machine that controls elements in the work cycle. In the past, the old techniques of 'time and motion study' and its later form 'work study' concentrated on the accurate measurement to seconds of every element of motion and activity within a process. These techniques are still used, but in many companies the emphasis has moved away from that approach to one of increasing the motivation of people. In progressive companies this is being done by enhancing the workers initiative, personal responsibilities and influence on their working practices and environment.

Planning decides in which sequence a product will be manufactured, tested and packaged. The sequence is normally documented on process planning layouts which may have a variety of similar names. Such documents list the processes in sequential order; the materials, tooling and gauging needed, plus the standard times for production. In some companies, planning departments within engineering may also load the jobs onto the production floor and progress them through the sequence. It is, however, more usual in larger companies to have a separate Production Planning and Control department reporting to the senior production manager to do this.

Cost Estimating

Businesses need to know the cost of the items produced. The estimating function may be carried out by a separate department or by manufacturing engineers working within another function. Estimating is in some cases located in the Production Planning function or Sales function because of the types of cost that can be required. These may be:

1. Estimated costs for a one-off or batch quantity for direct quotation to a potential customer on 'make to order' jobs. In this case the engineer's manufacturing cost will be incorporated in a quoted price that includes company overhead and profit provisions.
2. Standard costs for inclusion in a standard costing system for regular make-for-stock production. The standard cost will normally include departmental overheads of the producing unit.
3. Forecast standard costs for calculation of the financial benefits of capital investment projects in new production equipment or new product developments.

The final costs produced will include labour, materials and overhead costs. These issues are explored further in Chapter 7 under types of cost.

Tooling Services

Production operations usually need some form of jig to hold the workpiece and tools to fashion it. These can be provided by internal departments or purchased from jig and tool designers and manufacturers. If done in-house, the two functions of design and manufacture will be jig and tool design.

Jig and Tool Design

Jigs (and fixtures) are needed to hold components during processing and tooling is required to perform the actual process, e.g. cutting tools or brazing heads. These are normally designed by jig and tool specialists who may also design special-purpose machines built for a specific process. The introduction of computer numerical control (CNC) machines in many processes has reduced the need for jigs and fixtures for many old-style conventional operations. Jig and tool design or selection is incorporated in some of the CAM (computer-aided manufacturing) software packages for CNC programming. However, many companies sub-contract this function to specialist designers and toolmakers.

Toolroom

Many manufacturing companies have a toolroom to make and maintain production fixturing and tooling on site. However, toolrooms are costly in terms of toolmakers' wages and the expensive accurate equipment needed. Also toolmakers are a scarce resource in many areas. Consequently, other companies buy in their tooling from sub-contract toolmaking firms, retaining only a minimum emergency repair or tool sharpening service.

Toolstores

Toolstores are often the responsibility of the engineering function because of the engineering nature of their stock. Many companies have central toolstores for the holding and issue of permanent and consumable tooling. However, with the modern trend to cell manufacturing or 'factories within factories', some companies now hold all tooling in the place of use, under the responsibility of the cell's resident manufacturing engineer. The main concerns for toolstores are the cost, frequency of use and access to tooling. In the case of consumable tooling (e.g. drills, cutters, etc), the cost of frequent replacement warrants the use of a sophisticated inventory control system to keep the value of the base stock as low as operationally prudent.

Facilities Engineering

Common activities and titles used within this core function include:

- Facilities engineering
- Factory engineering
- Process engineering
- Plant engineering
- Maintenance

Figure 3.4 shows a typical organisation chart for a Facilities Engineering function. The distinction drawn between facilities and production equipment is that facilities can be present in the absence of a manufacturing capability. For example, a distribution warehouse will have heating, ventilating and electrical power equipment plus the necessary engineering support, but no design or manufacturing functions.

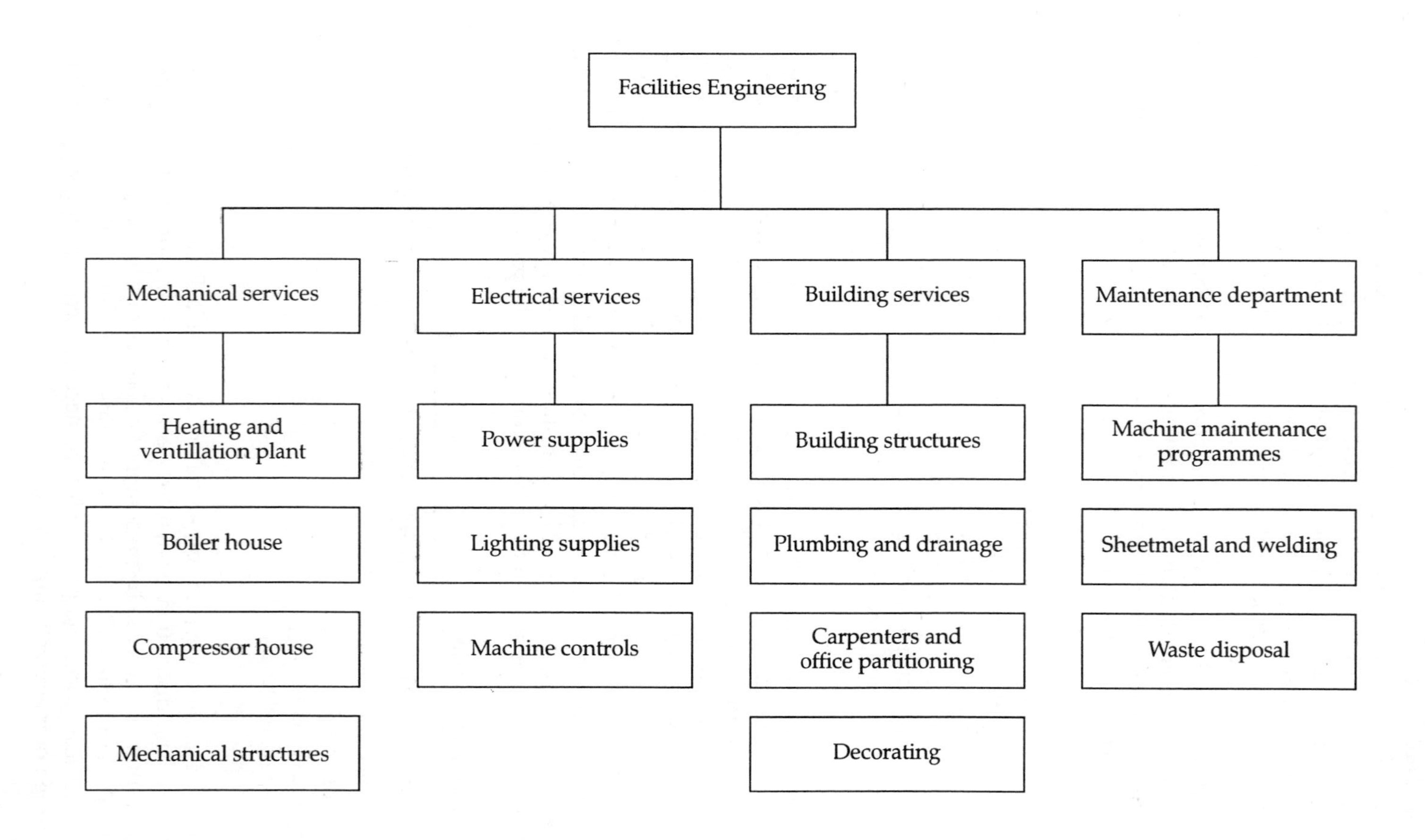

Figure 3.4 Facilities Engineering organisation chart

The function is also known as works, site or maintenance engineering. It is responsible for introducing and maintaining engineering services to the site for the provision of power, lighting, heating, ventilation, compressed air and general building maintenance. In industries producing capital goods such as cars through the assembly of components, the organisational distinction between manufacturing engineering and plant engineering may be quite clear. In other industries producing such items as bulk chemicals, the manufacturing processes are more aligned to large-scale plant engineering with their dependence on the use of heat, pressure and temperature rather than component production and assembly.

The specification and provision of site and energy facilities require knowledge that is usually distinct from the manufacturing technology employed – with the exception of process engineering and equipment such as plating and painting lines. Maintenance of all facilities and production equipment may be included in the facilities department's remit but it may be separate according to the volumes and balance of work. The staff on a large site will often be divided into the specialist areas that may include:

- Mechanical Services
  - Mechanical structures and welding
  - Heating and ventilating plant
  - Boiler house
  - Compressor house
- Electrical Services
  - Site power supplies
  - Machine controls (electrical and electronic)
- Building Services
  - Building structure
  - Plumbing
  - Carpenters
- Maintenance Services
  - Machinery and equipment routine maintenance

As stated previously, such specialisms may be amalgamated into fewer positions and also bought in by smaller companies. It is absolutely essential, however, whatever the organisational arrangements, to have a qualified electrician on site for safety reasons.

Quality Assurance and support

Common activities and titles used within this core function include:

- Quality Assurance (QA)
- Quality Control (QC)
- Quality Engineering (QE)
- Inspection

Figure 3.5 shows a possible organisation chart for a Quality Assurance function.

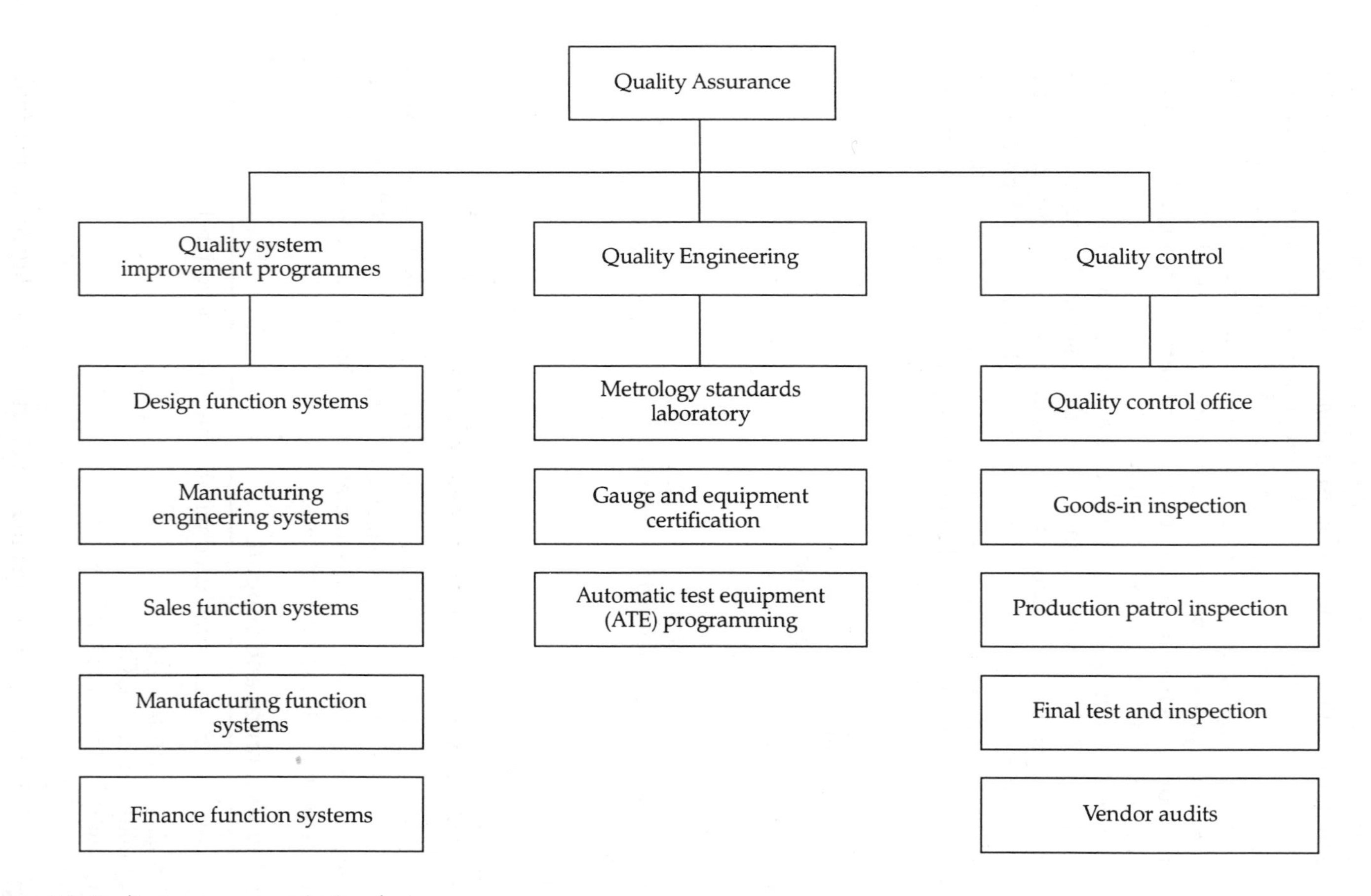

Figure 3.5 Quality Assurance organisation chart

Quality Assurance

Quality Assurance (QA) is a philosophy and control system applied at corporate level to enhance and maintain the quality performance of all the company's operations, not just the quality of production. In the UK the predominant QA systems in use are the world-wide ISO 9000 series of standards. ISO stands for International Standards Office which is the world body for co-ordinating standards. In the UK the standard is now designated (BS)(EN)ISO 9000 but was formerly known as BS5750. As another example, in Germany the standard is designated (DIN)(EN)ISO 9000. In Europe the system is now designated (EN)ISO 9000 but used to be known as EN 29000. These are now identical and are developments of earlier Ministry of Defence and NATO quality programmes which were imposed on military equipment manufacturers. These are now being applied to general commercial use as world markets become more competitive. All defence customers and an increasing number of other companies will no longer deal with a supplier that does not have the appropriate ISO or BS quality system in place.

With QA, the quality function has broadened from the physical inspection of production output to a system for applying a quality culture throughout a company. It must be pointed out, however, that these standards are procedural standards and NOT technical standards in the manner of BS, DIN, CEN, CENELEC and other national or industry engineering standards. This is borne out by the fact that many commercial businesses and public services such as hospitals, doctors' surgeries and swimming pools now have ISO 9000 series certifications.

The standard is divided into a series of parts, each dealing with a different scope of activity, for example:

- ISO 9001 – The design, development, production, installation and servicing of a product or service. Businesses that design and make a product would fall into this group.
- ISO 9002 – The production and installation of a product or service. Businesses that make a product to others' designs, i.e. sub-contractors would be covered in this section.
- ISO 9003 – The final inspection and test of a product or service. An inspection agency would qualify here.

Further sections of ISO 9000 covering other specific areas are in operation.

The major functions of QA are:

1. Introduce and establish (BS)(EN)ISO 9000 to the business, usually in conjunction with an external quality auditing organisation.
2. Carry out regular planned internal audits of all the business functions to confirm the existence of suitable and approved procedures of control, and confirm that they are being adhered to.
3. Liase with the external (BS)(EN)ISO 9000 auditing bodies who conduct bi-annual audits of the whole company. Approval through certification

by such an independent body is the best source of confidence that customers and suppliers can have that the company is conforming to the established practices. Note: internal auditing can include the physical testing of finished products.

Quality Control

Quality Control (QC) was the general term used to cover all quality activity before the advent of QA. In a more specific sense it is the day-to-day operation of the quality, inspection and statistical routines applied to physical production or services within the QA system. Best modern practice is to put more responsibility for the quality of produced items and their inspection with the producer and not an inspector from QC. This increases the sense of ownership and responsibility on the part of the producer and has been shown to improve quality. 'Inspectors' are focused on patrol and random inspection using statistical sampling methods to ensure conformance within the acceptable limits laid down for the design. Exceptions to inspection sampling are 100 per cent performance tests on sub-assemblies, final product and safety-critical items.

Inspection is usually the part of QC function concerned with the physical checking of goods and products at various stages within manufacturing. Commonly found activities include goods-in, in-process, final product and goods-out inspection.

Quality Engineering

Quality Engineering (QE) is a key activity that controls the use, condition and calibration of measurement equipment used in the production and inspection processes. The work is normally carried out in a standards laboratory (often temperature- and humidity-controlled) that has checking equipment calibrated to industry, national and international standards. This allows the quality of the product and the inspection equipment itself to be traced back directly to such standards. All inspection and test equipment has to undergo periodical checks to ensure this conformance. Accepted equipment is given a certification label in a prominent position. Rejected equipment must not be used and should be withdrawn to engineering for rectification. Another task of the quality engineering laboratories is to carry out investigations on non-conforming materials and on prototypes to establish the metallurgical, physical and chemical natures and performance of the materials.

3.4 The role of the engineer

The role of each individual engineer will be defined by both discipline and function and will vary with the gain in knowledge and experience. This gain brings a greater degree of responsibility even if it is outside the management structure.

Junior engineers

In this case the technical discipline of the engineer should be the primary consideration in matching engineer to role. However, the scope of that role will depend upon the objectives of the department and function. The objectives define the tasks to be performed and these will in turn develop an engineers' experience into a sub-specialism relative to the company's needs. This is likely to happen even if there is an initial period of 'doing the rounds' for an incoming engineer.

A junior engineer will normally be given relatively simple tasks to start with and will often work alongside a more experienced person who will act as mentor and informal supervisor. It is vital that junior engineers are given guidance and instruction and not just left to 'get on with it', as this creates greater technical risks for the company and a slower than desirable learning curve for the engineer.

A review scheme, including frequent mutual discussion and evaluation, should be used to help the junior engineer and the manager to achieve the best balance of induction, education and contribution. It is vital that young engineers are not regarded as persons that must 'serve their time' before making real progress. This is perhaps a common reaction from older, more senior staff who 'went through the mill'. Modern, highly competitive industry cannot afford to under-utilise trained resources and engineering managers must take this viewpoint. This is particularly true where new computer-aided technologies are used and the younger staff are more computer-literate than their senior colleagues.

Senior engineers

A senior engineer may often lead a small section or juniors in his/her specialist area. This may be reflected as a position on the management chain but is most often rewarded in salary levels. Senior engineers often find their specialist experience a blessing from the point of job attainment and satisfaction but a hindrance to branching into wider fields or management. Also they will be at the top of the engineers' seniority and salary scale and feel they have nowhere to go in terms of career progression and significant salary increases. This can cause resentment and frustration as senior engineers then see themselves by-passed by younger staff moving up the traditional hierarchical management ladder.

The modern trend to reduce the number of levels of management may at first be thought to exacerbate the situation by reducing the number of managers' posts available. However, the new group-centred philosophy of devolving more decision-making to the working group has been found to provide increased motivation and sense of responsibility. This applies to all but is particularly relevant to senior staff blocked from management positions.

The engineering manager

The engineering manager has a triple responsibility for:

1. The technical output of the department.
2. The technical competence of the engineering staff.
3. The efficient operation of the department through good management practices.

The balance of emphasis on technical versus operational performance will change with levels of management, the technical detail decreasing and operational management responsibilities increasing with the more senior posts.

3.5 Production

Production, that is the actual making of the product is neither an engineering nor a commercial function. It is, however, the core activity of any manufacturing company and the major functional customer of engineering activities such as design, manufacturing engineering and quality engineering. In addition, the production activities in an engineering company are by definition engineering in some form. In other manufacturing companies engineering services will provide and maintain production machinery, factory facilities and quality. It is important, therefore, that we examine the relationships of the engineering functions to production because these are two of the elements in the triangle of company activity shown in Figure 3.6. Figure 3.7 shows a typical production organisation structure.

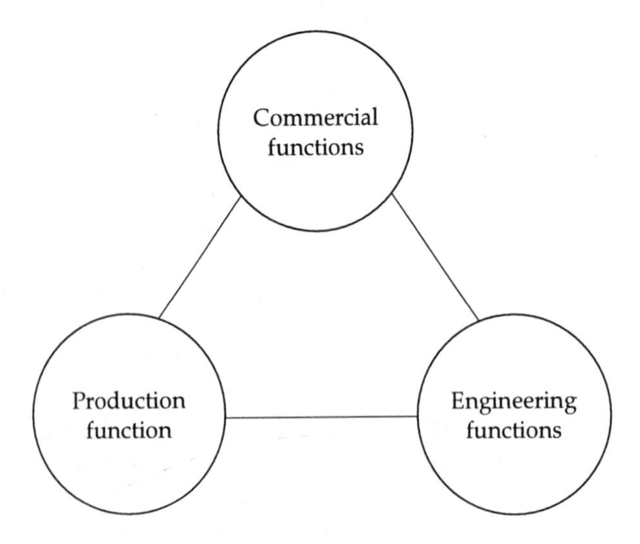

Figure 3.6 The functional triangle

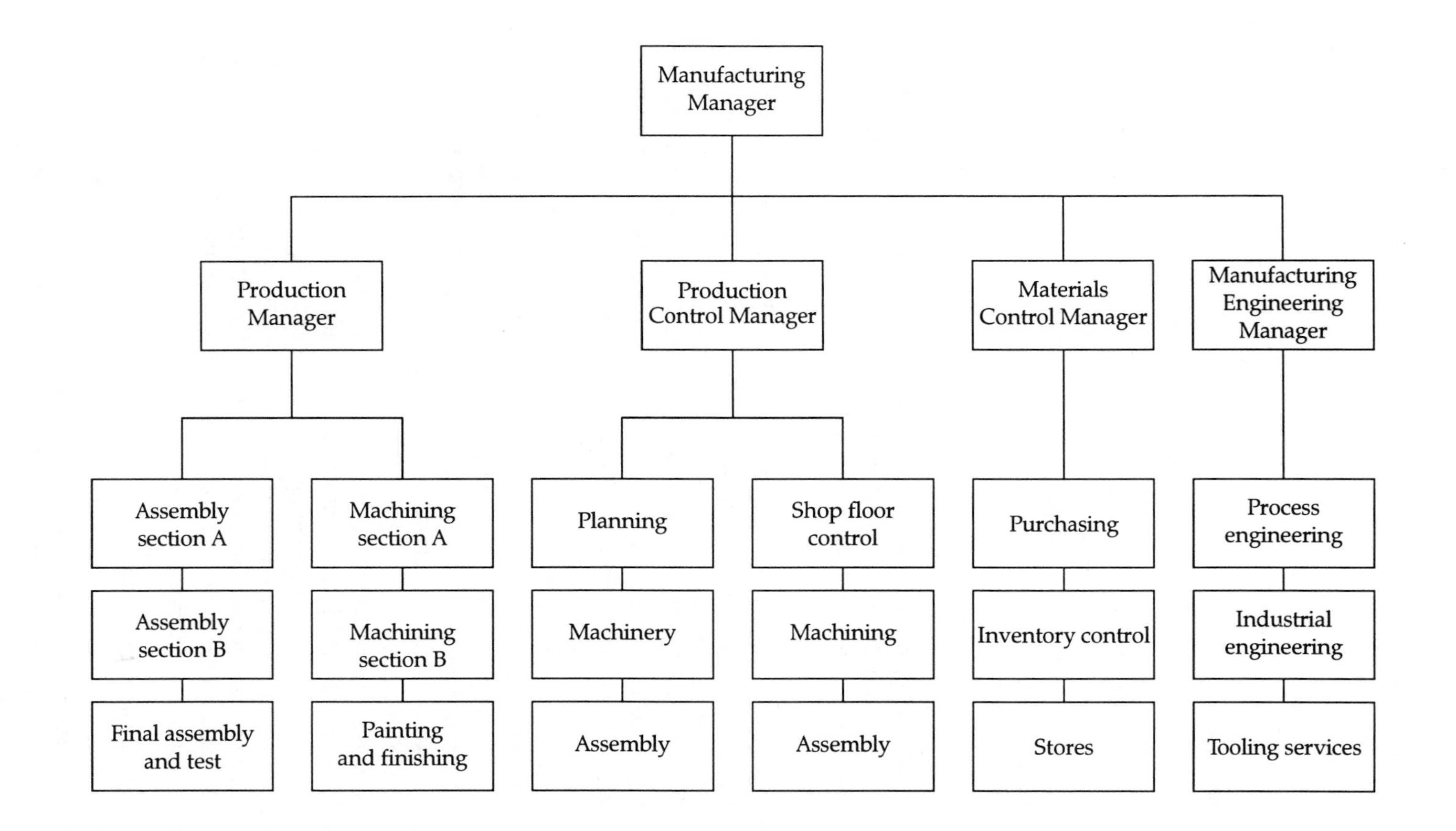

Figure 3.7 Manufacturing organisation chart

Types of Production

Production can be said to fall into three broad types:

1. Job or one-off production.
2. Batch production.
3. Continuous or flow production.

Job or one-off production

This is common with large, expensive or infrequently ordered products such as buildings, bridges, ships, power stations, or items of special equipment. These may be reproduced from a common design but it is more usual for the design to be unique or an extensive modification of a previous design. Because of the one-off quantities involved, unit costs are higher than could be achieved with greater volumes of production. Preliminary design work may have been done in order to quote or tender for the job but detailed design and production will only be undertaken against a definite order. That is, the item is 'made to order'. In both site work such as civil engineering and in the one-off production of items within a factory, the work will be organised as a project with a project management structure rather than a management structure for repetitive production.

Batch production

This is used where items are required in larger numbers but not to such an extent where they could be produced continuously. Standard designs are usually offered for sale although in many businesses, amended designs ('specials') will be offered if it is economically sensible. Production is organised to make batches of products on the basis of the orders received or a sales forecast. This allows the production processes to be automated to some extent, depending on the cost savings available in time savings versus the capital cost of the automation. All the manufacturing functions of the company will reflect the batch nature of the production processes. Product design, materials purchasing, manufacturing engineering, production planning and control, production and quality control will all be organised to reflect the batch production nature of the factory output.

Economies of scale can be achieved by selecting batch sizes for production that increase manufacturing efficiency but at the same time do not lead to the build-up of high levels of stock. This consideration is covered more fully in section 8. Another advantage of batch production is that it allows greater control for the purposes of scheduling, supervision, handling, quality control and cost control. This is achieved because each batch is separated both physically and in time and has a unique batch number which allows good control and traceability.

Continuous or flow production

Continuous or flow production is employed where the quantities are sufficient to require continuous or near continuous (i.e. very large batch) production. The large quantities normally justify capital expenditure on considerable automation of the production processes because of the low unit costs that can be achieved. Even though production may be continuous, the ordering and delivery of materials at the start of the production cycle and the packaging of the product into discrete quantities for shipment at the end, plus quality assurance requirements, impose a need for some system of control that gives quantity or batch identity and traceability. Typical industries using continuous flow are petrochemicals, food, drink, consumer durables (cars, TVs, refrigerators) and the even higher volume production of components for them. For example, one TV needs hundreds of electronic and electrical components produced for it.

Modes of production

Production can be achieved by various physical layouts and types of system control of the production processes. In the case of continuous production which we have just discussed, the process equipment is sited in a continuous layout sequence and the product automatically flows from one process to the next. The process equipment is dedicated to that product and it would be impossible to maintain economically sound continuous production without having an uninterrupted passage of the product from one process to the next.

The commonly used modes of production are:

1. Dedicated lines
2. Process-centred production
3. Cell manufacturing
4. Just in time (JIT)

Dedicated lines

As with continuous production, process equipment can be dedicated to one product but this is undesirable in batch production because there would be a large amount of under-utilised process equipment where the batch quantities would not use the time available in the working day.

Process-centred production

In this arrangement, all similar process equipment is grouped together. For example, in the machine shop, all lathes would be grouped together in a turning section and all drilling machines would be in a drilling section. The process equipment is not dedicated to one product and the various products are moved from equipment section to section as they follow their particular process sequence. The layout has the advantage of concentrating human skills in one area but has the great disadvantages of causing:

(a) a large amount of time-consuming product transportation between sections;
(b) long periods of time (called lead time) between the first process being started and the last one finished.

Notwithstanding these disadvantages, the process-centred organisation with its concentration upon the craft skills of workers has been the traditional layout of the majority of engineering and other manufacturing factories until recent times.

Cell manufacturing

In this type of layout all the process equipment needed to produce a product or sub-assembly is grouped in one area. For example, the lathes, drills, assembly benches and testing of a pump assembly would be arranged into a production cell which concentrated on that product. The equipment is, therefore, dedicated as in continuous production but does not have the high volumes of output needed to justify complete automated lines. The system has the advantages of cutting out inter-process travelling time, reducing lead times, keeping better control of the product between processes and maintaining a greater interest of the staff in the output of the cell as their end product; it is not just a component that disappears into an assembly somewhere else.

However, there are economic and space constraints on providing a complete process set-up for each cell. For example, it would be very costly and space-consuming for every cell to have its own paint spraying and metal plating process. The cost of providing the special tanks and statutory ventilation systems would be uneconomic for most businesses.

Just in Time (JIT)

Introduced from Japan, this is a development of the cell principle aimed at keeping the levels of work-in-progress (WIP) to an absolute minimum. Low WIP levels are achieved by having the cell team of staff decide when they want components supplied or made by themselves, rather than Production Control pre-planning and scheduling all the batches and their sizes. The ideal batch size for JIT is said to be one, so that for example, only one cup handle is produced when one cup is needed, rather than pre-planning the production of 100 handles for use in the future. In practice, there are physical capacity and economic constraints placed on this 'one for one' ideal system, but nevertheless by aiming at the JIT principle, great savings in WIP levels and lower lead times for production start to finish are achieved.

Figure 3.8 shows layouts for the alternative modes of production.

Production Control and systems

Production is a dynamic activity involving the performance of the many tasks needed to convert a series of raw materials and purchased components into a finished product. The size of this undertaking is possibly best illustrated by the fact that in the majority of manufacturing companies the production function employs more staff than any other and this commonly ranges from 40 per cent to 60 per cent of the total staff. The activities of so many people need to be subject to formal control systems if complete confusion and inefficiency are to be avoided. The task of the Production Control function and its allied systems is to provide and operate the formal

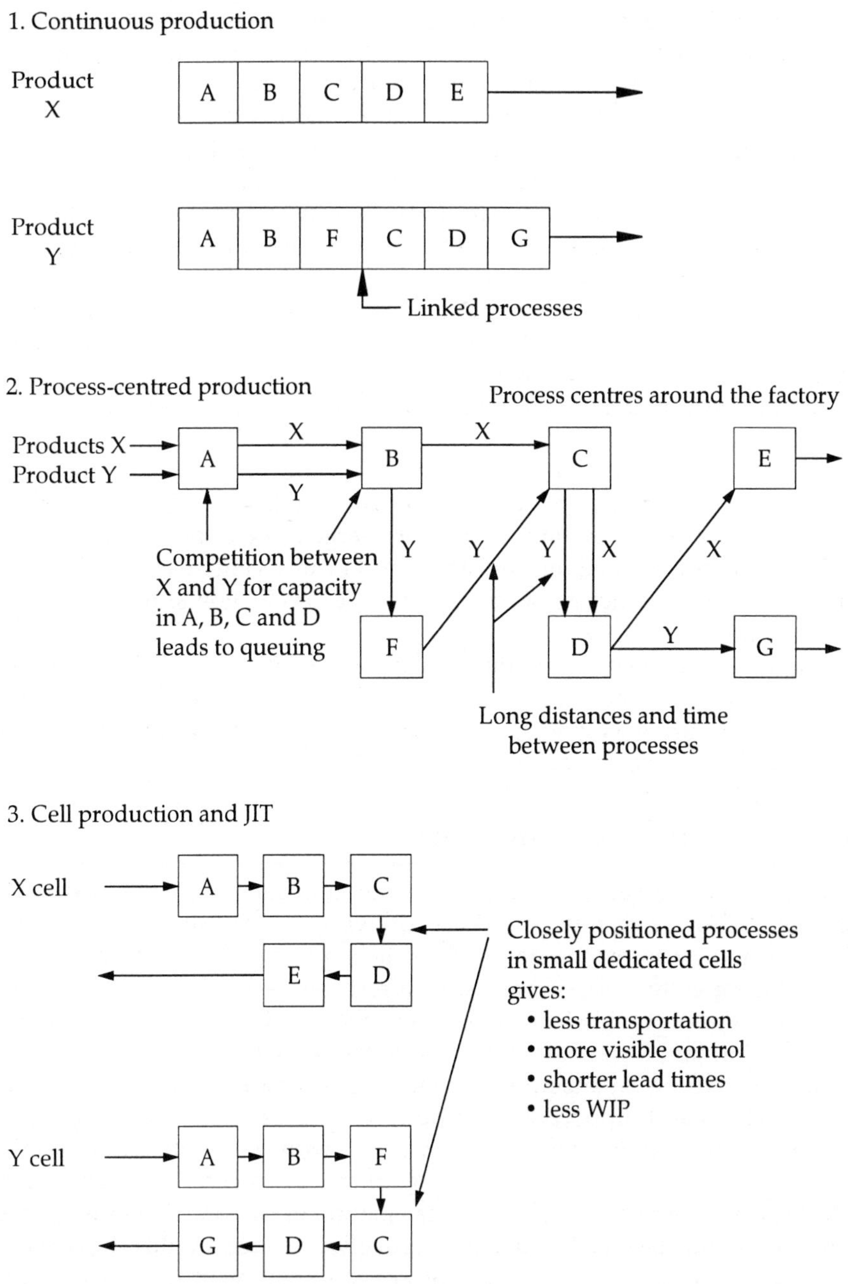

1. Continuous production
Product X
A B C D E
Product Y
A B F C D G
Linked processes
2. Process-centred production
Process centres around the factory
Products X
Product Y
A
X
Y
B
X
C
E
Competition between X and Y for capacity in A, B, C and D leads to queuing
Y
Y
Y
X
X
F
D
Y
G
Long distances and time between processes
3. Cell production and JIT
X cell
A B C
E D
Closely positioned processes in small dedicated cells gives:
• less transportation
• more visible control
• shorter lead times
• less WIP
Y cell
A B F
G D C

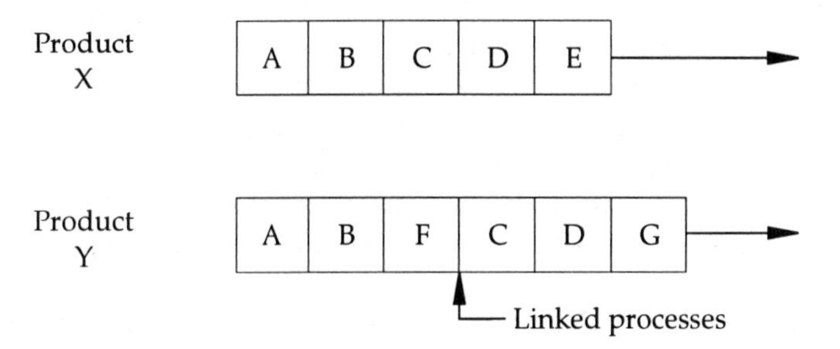

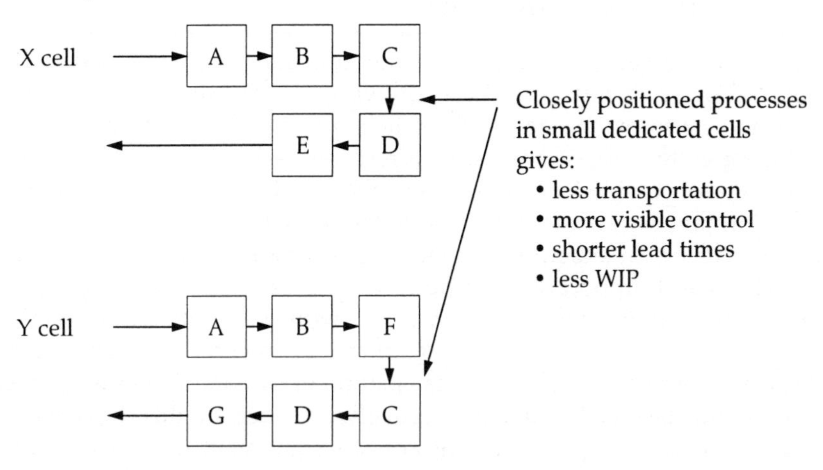

Figure 3.8 Modes of production
Note that products X and Y follow the same process sequence through three different process layouts or modes of production. (Courtesy of Butterworth Heinemann).

planning, scheduling and control of materials and production operations throughout the entire production sequence from raw materials to finished product despatch. Figure 3.9 shows the main elements of a production and materials control system and the flows of data within it.

The starting-point for the flow of information is the production design and details which provide basic data of what needs to be made and what form it shall take in order to produce the final product. Parts lists produced by the Drawing Office are normally converted into Bills of Material (BOM) which organise the materials into the groups that are most logical for production operations as opposed to the product design. Information from sales forecasts and stock levels are used to produce a Master Production Schedule (MPS). This and the bills of material are then used to drive the detailed planning of the many production operations involved.

It must be appreciated here that Production Control systems are not dealing with just the production of a quantity of one product over a time period but usually are concerned with the parallel production of many products in varying quantities and to different delivery dates. In the past, this has necessitated creating systems involving a large number of clerical planning and scheduling calculations and tracking but in recent years this has been aided by the introduction of computer-based production and material control systems.

Engineers, particularly manufacturing engineers, are likely to become involved in production control issues in their work and it is quite common for manufacturing engineers to move into positions in Production and Material Control departments.

Engineering services to production

Production is served by the engineering functions in many ways. The design and drawing office produces drawings, parts lists, material specifications and performance standards to be supplied either to a central data store for repetitive use, or are issued afresh for each job. Trouble-shooting support is also provided in cases where Quality Control cannot resolve a 'non-conformance to standard' in production output.

Manufacturing engineering supplies the technical expertise needed for the provision and support of the production processes and equipment used therein. Industrial engineering is concerned with the creation and issue of process descriptions and methods working instructions (usually called worksheets), routeing data and the calculation of standard times and costs. Shop floor and factory layouts are normally provided by this department. Tooling and special-purpose machines or test equipment may be designed and made in house. In many companies the function is also concerned with production control and inventory control systems as part of services needed by production. Production engineers (sometimes known as line engineers) are often resident within a production area to provide on the spot day-to-day process trouble-shooting.

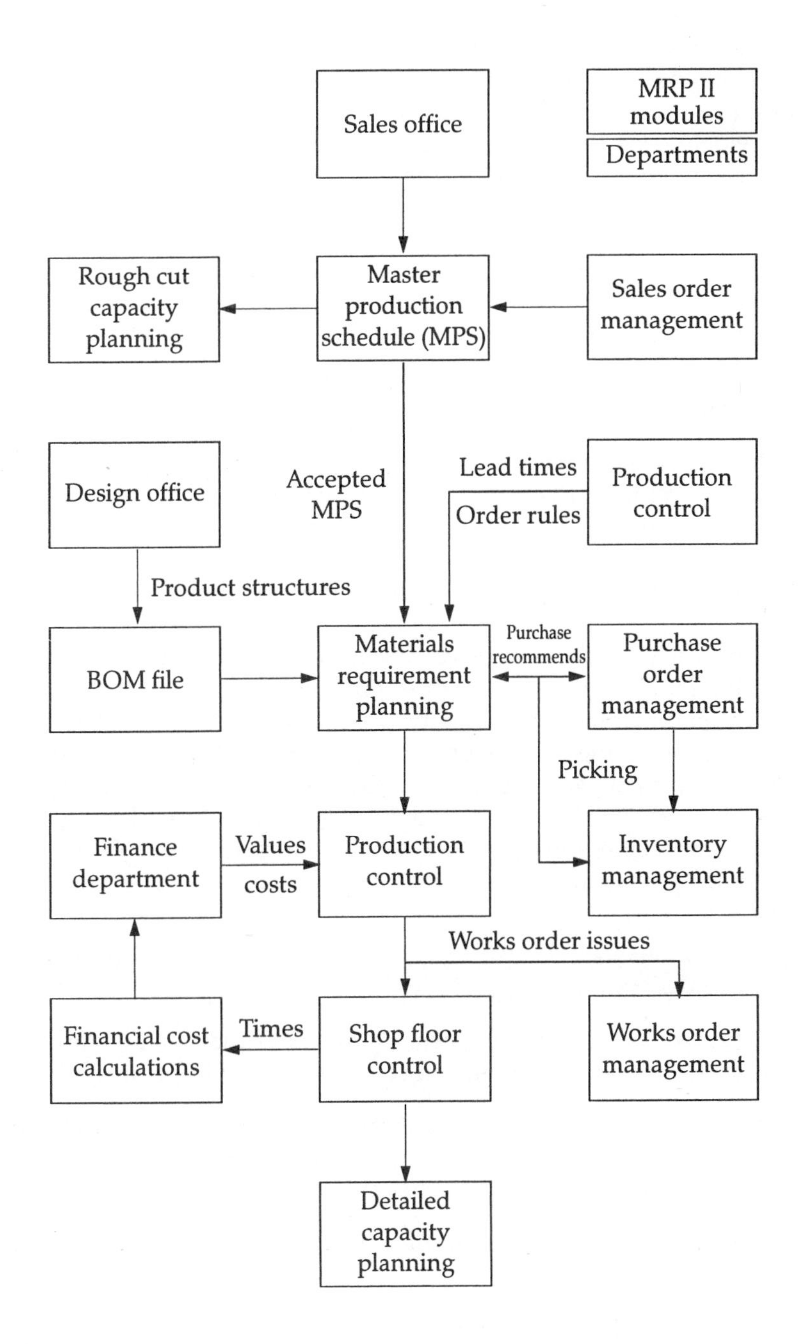

Figure 3.9 Main elements of a production and materials control system (Courtesy of Butterworth)

Manufacturing engineering has been shown as reporting to a company-wide engineering function for the sake of explaining the relationships of engineering practice. In many companies, however, manufacturing engineering is part of the the production function because of its very close association with production activity.

Quality Assurance (QA) has expanded from the inspection of production output to a company-wide quality philosophy and control system. The system widely used these days is the ISO 9000 series. QA will routinely audit the production administration routeing and authorisation procedures for conformance to the standard. Quality Control (QC) audits and measures production output and incoming materials by applying statistical methods to sample inspections of components and materials. The exceptions to sample inspection are normally the testing of the final product, sub-assemblies or safety-critical items which will be 100 per cent tested. QC will decide on the acceptability or not of production output that is border-line to the standards required.

Quality engineering is concerned with the maintenance of technical standards within all inspection equipment. A metrology section will maintain reference standards traceable to industry or national ones and there will be a gauge and inspection equipment certification routine. Equipment that is not certified must not be used.

Facilities engineering provides both direct and background services to production. Electrical power, compressed air, water and drainage are supplied to machines and processes. Special areas such as clean rooms will be provided for electronics and high technology work where the control of air quality, temperature and humidity is critical. In other areas with potential health hazards from fumes or dust such as paint spraying, high output extraction systems will be installed. General background lighting, heating and ventilation will be provided to the factory and offices. Factory buildings and access facilities must also be provided and maintained. Planned routine maintenance of all these facilities will be carried out against a schedule.

3.6 Exercise 3

1. Describe the four core engineering functions that can be said to encompass the majority of engineering activities found in business.

2. Are engineers restricted by their discipline field (i.e. electrical, mechanical or manufacturing engineer) to a particular core function? Give examples to demonstrate your conclusion.

3. Outline the main constituent functions of Product Engineering and their primary tasks.

4. Outline the main constituent functions of Manufacturing Engineering and their primary tasks.

5. Outline the main constituent functions of Quality Assurance and their primary tasks.

6. Outline the main constituent functions of Facilities Engineering and their primary tasks.

7. Which two of the four core functions are the most likely to be combined into one function in a small company? Give your reasons why this should be so.

8. Explain the work of the following departments:

   (a) R&D
   (b) Product Design
   (c) Drawing Office

9. Explain the work of the following departments:

   (a) Process Engineering
   (b) Industrial Engineering
   (c) Line Engineering

10. Explain the work of the following departments:

    (a) Quality Assurance
    (b) Quality Control
    (c) Quality Engineering

11. Draw an organisation chart to show how the functions mentioned in questions 8, 9, and 10 might fit into the organisation of a medium-sized engineering company, together with Sales and Manufacturing.

12. Consider three types of production and which products and industries might be expected to feature these types.

13. Consider three modes of production and which products and industries might be expected to employ these modes.

14. What are the major advantages of the cell production mode over the process-centred mode?

4

The commercial functions

4.1 Introduction

All activities and functions within businesses have a commercial objective in the sense that they contribute in some way to the sale of a product or service for profit. Indeed, the objective of this module is to make engineers aware of this input and its connection to the commercial world.

Within this text the term 'commercial function' is used to describe those functions that are not predominantly concerned with the actual production of a product in the factory, or providing the engineering and technical services needed to do so. Thus, functions such as Marketing, Sales, Distribution, Purchasing and Finance are considered to be commercial functions and we will now look at these in turn.

4.2 Marketing function

Marketing should not be confused with sales in the mind of the engineer because they are two distinct functions. Marketing is concerned with the issues of which products shall be sold, where they are sold and how the company will compete in the market place. Sales is concerned with direct contact with potential and existing customers to gain an order. Figure 4.1 shows some possible organisation structures for the marketing and sales functions. Figure 4.1 shows them combined in one core function reporting to the. Managing Director (MD). Figure 4.2 shows them as separate functions. There is no general rule that can be applied to where the two structures may be split, but the larger the business, then the greater likelihood of the separation of functions and the smaller the business the greater the likelihood of amalgamation.

Marketing activities

The activities of a marketing function will include:

1. Market research, which is used to gain customer reaction to existing

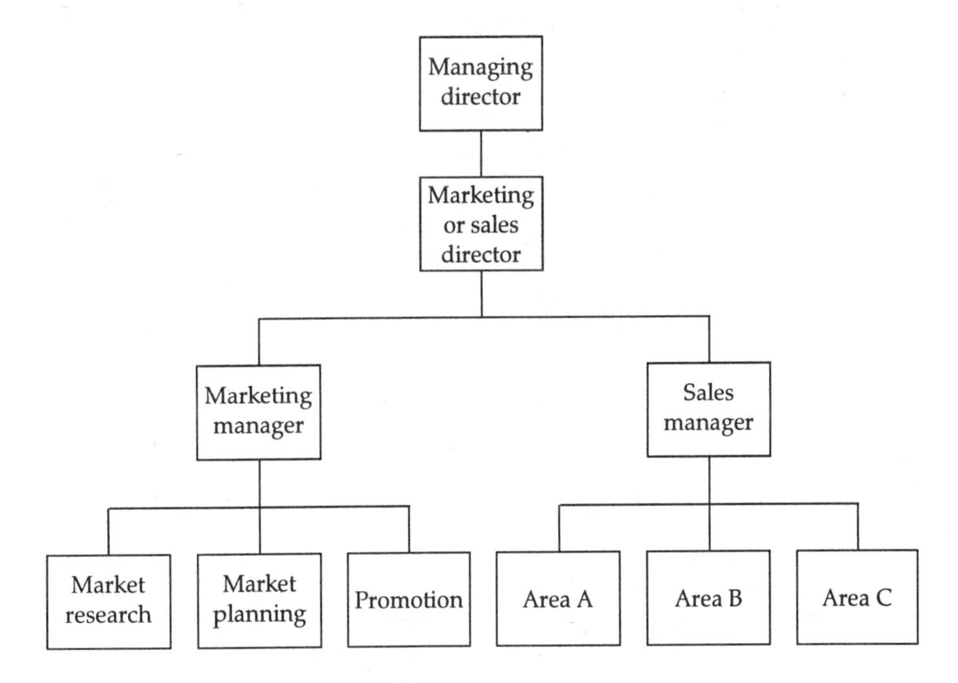

Figure 4.1 Combined marketing and sales organisation

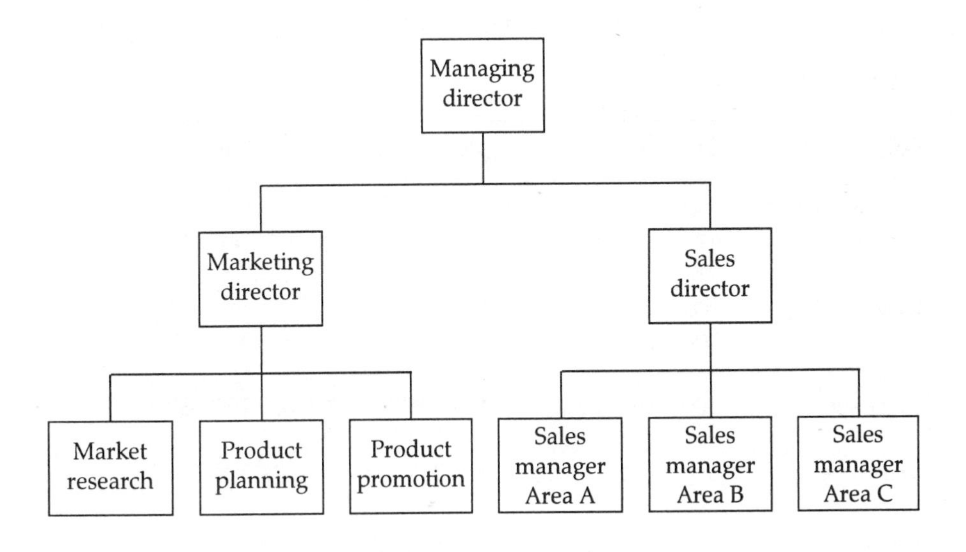

Figure 4.2 Separate marketing and sales organisation

products (own and competitors') and to what the future requirements of the market might be.

2. Deciding which sectors of the market to concentrate on. For domestic products this will involve deciding which socio-economic category is to be targeted. For industrial products, which industry or part of an industry should be targeted.
3. Deciding which products will be sold and the balance between them with regard to features, price and quantities to provide a comprehensive range of choice to the customer. Taking cars as an example, manufacturers may offer a wide product range based on four parameters such as:
   (a) basic size;
   (b) body style;
   (c) engine size;
   (d) market sector and image.

   Table 4.1 shows a typical range of cars available that could be marketed from combining the parameters. It can be seen that by taking permutations of the parameters, up to 24 different models can be offered for sale without making 24 completely different sets of car parts. Many of the sub-assemblies and detailed parts like steering, suspensions, gearboxes, braking systems, and seats will be common to many variants. Also, a basic body shape can be used to produce different styles and allow annual upgrades by the alteration of front grilles, side panels and rear body parts. This marketing strategy enjoys three great advantages:
   (a) a wide range of choice for the customer;
   (b) economic production of high volumes of common parts;
   (c) inexpensive annual detail changes to keep market interest.
4. Pricing policy is a key consideration. Price must take account of the competitiveness of the market in order to obtain the targeted sales volume

Table 4.1 Product range for cars

Basic body size	Small	Medium	Large
Body styles offered			
Hatchback	Yes	Yes	Yes
Saloon	Yes	Yes	Yes
Estate	No	Yes	Yes
Sporty/GTI	Yes	Yes	No
Engine size range cc	1000 and 1300	1300, 1600 and 2000	2000–3000
Models offered	$3 \times 2 = 6$	$4 \times 3 = 12$	$3 \times 2 = 6$
Total offered = 24			
Overall images	Economical and fun car	Family and reliable	Executive and luxury

but, on the other hand, it must also fully recover the profit targets and total costs of the product including all overheads. Price discounting is common in the more competitive markets as is discounting for high quantity orders or long-term contracts. Prices will often be set below the competition's prices in order to establish or increase a share in the market volumes available (known as market share e.g. 30 per cent). This is sometimes done to generate additional sales as part of a long-term marketing strategy, even if it reduces the initial profit.

5. Product promotion is also an important activity. With consumer products it is normally carried out via advertisements in the media (i.e. newspapers and TV) and in retail outlets. Promotion for commercial or industrial products is usually done through advertisements in trade magazines, or promotional seminars, exhibitions and conferences. The function may be called Publicity or Advertising in various businesses.

Market strategy

Two major elements are involved in deciding the actual strategies to follow which are:

1. the product;
2. the market.

By combining the existing situation on these with possible developments, four marketing strategies can be constructed.

- Increase market share with existing products in existing markets.
- Develop new market areas with existing products.
- Introduce new products into existing markets.
- Develop new products for new markets.

The level of risk involved varies considerably and increases from the first to last listed programme because larger amounts of money are needed to finance entirely new projects and there is a greater element of uncertainty of the result.

A business with a wide range of products and markets is likely to be carrying out the first three strategies with respect to a number of products at the same time. However, developing new products for new markets involves the very highest of risks and will be the subject of much prior investigation before a decision is taken at boardroom level.

Engineering support for marketing

The main link between engineering and marketing is in the planning of new products or development of existing ones. Marketing will provide R&D or Product Design with some parameters or a design envelope to work within. In return, design feasibility studies or proposed outline designs will be submitted to Marketing or a company New Product Committee for a decision on undertaking of further work.

Summary

The aims of the marketing function are to recommend to management which products should be marketed and what marketing strategies should be employed. The two major outputs of this are to influence the direction of R&D work on new products and to provide the direction of the activity for the sales functions.

4.3 Sales function

General structures and activities

Sales is concerned with the direct contact with potential customers for the presentation and sale of the product. The sales function is often divided into territorial areas either nationally or internationally or, alternatively, by product type sales forces selling to the whole market. This is often the case with technical products that need a good deal of specialist knowledge of the performance and possible applications of the product. The major activities of a sales function are:

1. The organisation of sales areas and salespersons into appropriate configurations for the markets being addressed.
2. The production of a sales forecast. This is an extremely important document because the production schedule and hence the material purchasing schedule will be based upon it.
3. The recruitment and training of salesmen into the company sales philosophy and practices, and providing a sufficient knowledge of the product and its competitors.
4. The maintenance of good relations with existing customers. It is generally easier and cheaper to keep a customer than find a new one.
5. The establishment of contacts with potential new customers.
6. The agreement of sales at agreed prices, allowable discounts and acceptable delivery times and, where applicable, agreed contract conditions. Most companies have standard terms of agreement on the back of their quotations and acceptance of orders. If the customer does not challenge these they automatically become the agreed conditions of sale.

Figures 4.3 to 4.5 show some possible organisation structures of the Sales function and its customer service activities. Figure 4.3 shows a sales function organised by sales areas in which the salespersons sell all the company products. This is applicable for many consumer or general engineering items that are being sold from an illustrated catalogue of standard products. Figure 4.4 shows an organisation split by product type. This is often applicable with complex engineering or scientific products where the salesperson has to enter into detailed discussions on product performance in new or challenging environments. Figure 4.5 shows a combination of

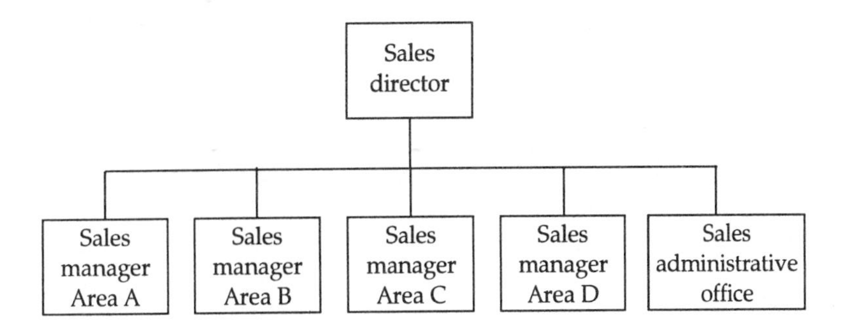

Figure 4.3 Sales function organised by sales area

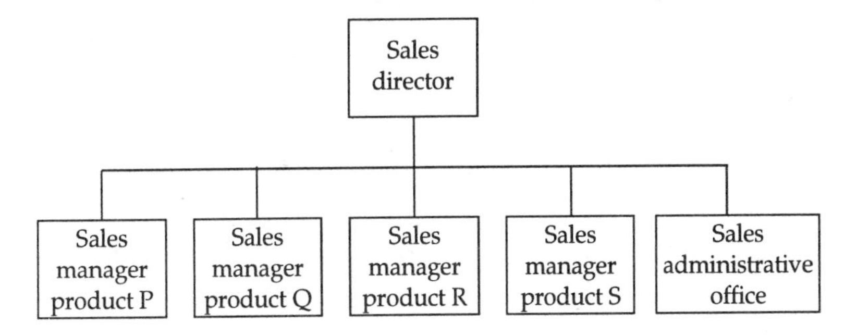

Figure 4.4 Sales function organised by product type

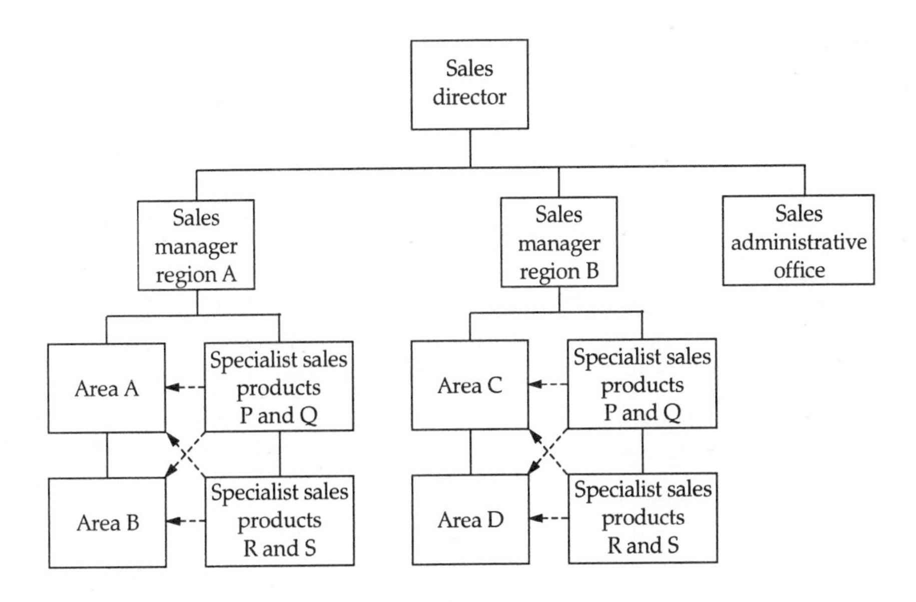

Figure 4.5 Sales function organised by area plus specialist support

Table 4.2 Annual Sales Forecast for ABC Co. Ltd

Sales area/product			Q1	Q2	Q3	Q4	Total
Units of sale							
Area North							
	Product A		10000	10000	9000	12000	41000
	Product B		5000	5200	4000	6000	20200
	Product C		1000	1100	1100	1400	4600
	Area total		16000	16300	14100	19400	65800
Area South							
	Product A		8000	8100	7500	8500	32100
	Product B		4000	4100	3800	5000	16900
	Product C		900	950	925	1000	3775
	Area total		12900	13150	12225	14500	52775
Total units							
	Product A		18000	18100	16500	20500	73100
	Product B		9000	9300	7800	11000	37100
	Product C		1900	2050	2025	2400	8375
Grand total			28900	29450	26325	33900	118575
Sales value (£)							
Area North		Unit price £					
	Product A	5.00	50000	50000	45000	60000	205000
	Product B	10.00	50000	52000	40000	60000	202000
	Product C	20.00	20000	22000	22000	28000	92000
	Area total		120000	124000	107000	148000	499000
Area South							
	Product A	5.00	40000	40500	37500	42500	160500
	Product B	10.00	40000	41000	38000	50000	169000
	Product C	20.00	18000	19000	18500	20000	75500
	Area total		98000	100500	94000	112500	405000
Totals	Sales value £						
	Product A		90000	90500	82500	102500	365500
	Product B		90000	93000	78000	110000	371000
	Product C		38000	41000	40500	48000	167500
Grand total value			218000	224500	201000	260500	904000

Table 4.3 Analysis of sales forecast

Model	Unit price £	Value p.a. £	% value	Units p.a.	% units	Comment
A	5.00	365500	40.0	73100	61.6	high volume/low price
B	10.00	371000	41.0	37100	31.3	mid-volume/mid-price
C	20.00	167500	19.0	8375	7.1	low volume/high price
Total		904000	100.0	118575	100.0	

both structures in order to cater for a situation where there are large sales areas to be covered with a mixture of standard and complex products. In this case territorial salespersons will be the local initial contact but central applications engineering function will provide specialist support when needed. Alternatively, product specialists may be present in each major sales area, particularly if they are national areas, involving different languages.

Example of Sales structure, issues and forecast

The sales forecast shown in Table 4.2 is for an engineered product that is sold in three versions or models – standard (A), de luxe (B) and specialist (C). We will now examine what can be learned about the company's business and sales functions from these without actually knowing what the product is. We can draw the following observations from the data which are also illustrated in the graphs in Figure 4.6, total units and Figure 4.7 sales revenue by product. Table 4.3 shows an analysis of the sales forecast.

Product range

Models A, B and C show a spread of the product offering across the potential market range from low price/high volume to high price/low volume versions. This enables the company to capture a share of the market at all levels and enhance product loyalty when buyers decide to move 'up-market' to a better version. This is a very important feature of any marketing and sales strategy, to put it bluntly, once you have a satisfied customer they are likely to stay loyal to the brand in a move to a better product. If the company did not offer models B and C, initial purchasers of A would be forced to buy from competitors when making a change to a product of type B or C.

Sales demand pattern

The forecast shows that the quarterly demand fluctuates significantly on models A and B. It stays fairly level for quarters 1 and 2 but then dips significantly in quarter 3 before rising to an annual peak in quarter 4. What does this immediately tell us? It indicates that the product is subject to seasonal demand fluctuations with quite a large percentage variation. On the other hand, model C has a fairly level demand with no seasonal dip, but this is to be more expected with an expensive specialist product unless it is suitable as

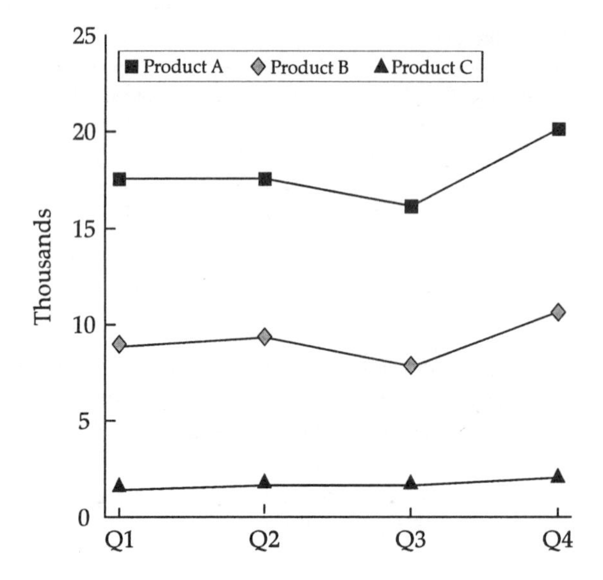

Figure 4.6 Annual sales forecast for ABC Co. Ltd: total units

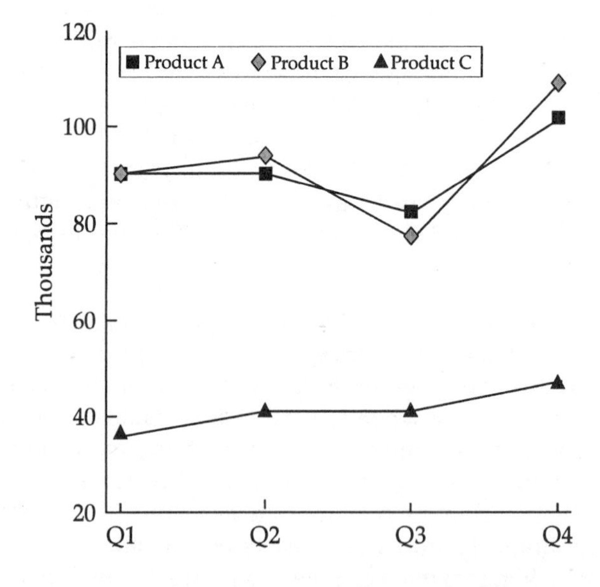

Figure 4.7 Annual sales forecast for ABC Co. Ltd: sales revenue by product

a personal present, in which case a rise in sales could be expected in the fourth quarter, i.e. for Christmas presents.

The effect of price on demand

The lowest price model A is evidently expected to sell twice as many as B because it meets the minimum customer expectation for that type of product at a competitively low price of £5. Model B may sell only half as many at a price of £10. However, if we compare the annual revenues for products A and B as shown in Table 4.3 we see that the total revenue is in fact highest for product B at £371 000 against £365 500 for model A. This would indicate that it might be better to concentrate more on product B (more revenue for less work), but there are two additional factors to consider.

First, although some people buying the £5 model might be prepared to pay £10, it is likely to be a fairly small number, thus the total revenues of the company from selling only B in preference to A would almost certainly fall.

Second we cannot make a sound judgement at this stage because we do not know the cost and profit margins for products A and B, and these would need to be included in any calculations regarding a possible change in sales policy.

Sales performance levels in the areas

Comparing the figures in Table 4.2 it can be seen that Area North is selling more than Area South on all three models. This situation needs investigation by the management to find which of several possible factors is causing this. The factors could be:

- The areas are not evenly sized. South may have a smaller population or much larger geographical area to cover.
- There is more competition in South. There may be more competitors, keener prices or better products to compete with.
- Area management is poorer than in North.
- Sales staff may be less well trained or able.
- The support given to South by head office in sales campaigns and marketing may not be so effective.

Whatever the reasons, the sales forecast combined with last year's actual sales give senior management a clear indication that the situation needs investigation.

Engineering support for the Sales function

We have already examined the subject to some extent in Chapter 3 with mention of the sales support by Product Engineering. The other core engineering function most likely to have a continuing dialogue with sales is the Quality function. We will now look at their contribution to sales.

Product engineering

Support will be given to sales in the form of expertise and data on the performance, technical details, repairs, spares and servicing of products which are being quoted for, being designed, being manufactured or already in the customer's possession or about to be sold. Most businesses that design and manufacture a range of products will offer a certain number as standard 'off the shelf' items. These will in most cases be the particular models and sizes within the range that have a regular sales demand. They may or may not be made for stock but the important point here is that they are of a standard design for which the salesperson can quote a standard price from a catalogue and price list. With industrial goods, the standard designs offered do not provide the ideal solution for all possible customer applications and the customer may need a special design. This is the point where a sales person's engineering knowledge does not allow them to fully specify the design departures needed and it is here that applications engineers or sales support engineers become necessary.

They will be product engineers with a good knowledge of the possibilities and limitations of the standard design and what design amendments might be introduced and be possible to manufacture and at what cost. It is at this point that manufacturing engineers may be involved. Well-organised companies have a formal procedure for handling such design departures or 'specials' because experience has shown that much money and time can be lost in pursuing this type of project, only for the customer to reject the solution because of the heavy premium it often places on the standard price. Poorly organised companies 'look at it to see if it can be done' without realising the full extent of the additional internal costs incurred if the exercise is to be technically sound and profitable.

Quality Assurance

QA will be involved with sales primarily to ensure that what the customer specified is in fact delivered. Parts of ISO 9000 and BS 5750 are concerned with Contract Review and Design Control. QC will be involved with any customer complaints and technical problems in service. A reporting system with documentation is normally used to ensure that all the relevant information gets back to the appropriate functions whether they are sales, design, manufacturing or distribution.

Manufacturing engineering

Manufacturing engineering can be involved with customers through the provision of tooling or service aids as part of a contract. Facilities Engineering may provide the site services to a regional Sales Office, particularly if it is attached to a company-owned distribution warehouse. Alternatively, sales operations may be in rented office space.

Summary

The sales function is the most obvious and direct contact of the business with its customers. As such, it always commands a high position in the organisation structure, reporting at board level either under the name of Sales or sometimes as part of a Marketing function. The importance of the Sales function can be demonstrated by the fact that actual sales will determine the revenues of the business, whereas most other functions incur the costs of the business. This one fact more than any other demonstrates the financial risks that are built into any business and all its constituent operation. The sales forecast is the primary document for the planning of all other functions and activities. Engineering support is given to sales in support of potential orders and in after-sales technical trouble-shooting.

4.4 Distribution function

Types of Distribution

Distribution is concerned with the efficient delivery of products from the source (normally a factory) to the point of sale. It can take many forms but three of the most widely used are:

1. Networked delivery to regional warehouses, from which deliveries to retailers or customers are made. This is usual with most consumer products having high volumes of production and many customers broadly dispersed around the market. In this case products are made to a production schedule derived from a marketing plan or sales forecast. Finished goods may be delivered from several factories making different items to distribution warehouses placed at convenient locations to serve the regional market areas. Here, the items are stored in bulk, often on pallets in high-rise racking. Individual items are then picked to satisfy the product mix in orders from local sales offices, retailers or direct customers.
2. Direct delivery from the factory to the customer. Companies producing small numbers of high cost industrial items will usually ship direct to the customer. However, even with this category of product, export orders may warrant having a combined small reception warehouse and service centre in the importing country.
3. Direct delivery from factory to retail outlet belonging to the manufacturer. Even so this will still require some order picking from a finished goods store within the factory.

Figure 4.8 illustrates the different distribution methods.

Functional position of Distribution

Distribution can be large enough to be a stand-alone function or may report to the marketing/sales organisation. In companies with a large distribution network involving regional warehouses and the mass trucking of product by container lorries, the function may be considered important enough to report at board level. In other businesses it may report to the sales function which is its main 'customer'. Alternatively, in other companies production and distribution are combined into an 'Operations function'. At the other end of the scale, distribution in a small company may be the occasional delivery of a product in the owner's car.

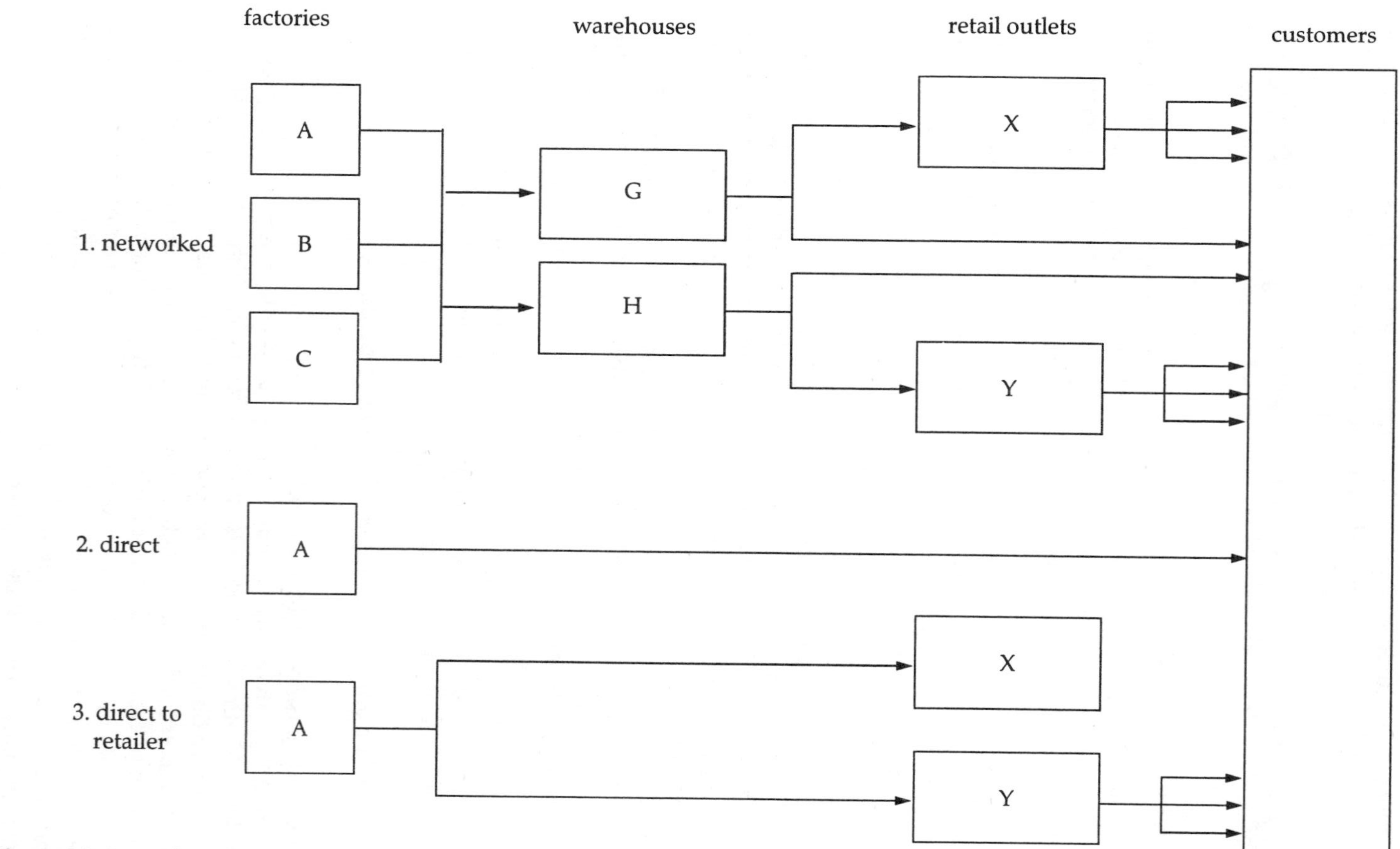

Figure 4.8 Types of Distribution

Distribution of Imports and Exports

The organisation and handling of goods for import and export are a complex procedure and worthy of mention here. The physical arrangements for getting goods off lorries and onto ships and aeroplanes are difficult. Added to this, the requirements of the national Customs and Excise organisations create a lot of regulations and controlling documentation to ensure that:

- only allowable goods are imported or exported. There are many national restrictions on engineering and scientific goods for military/security reasons and on foodstuffs and chemicals for agricultural/health protection.
- the excise duties levied by the importing nation are correctly applied and noted for future payment by the companies.

These issues are so complicated and time-consuming that any organisation moving even moderate amounts of goods internationally will have a Shipping Office or Import/Export Department within the Distribution function to handle the volume of administrative work involved. Even with such a department in place, the actual handling of goods and documentation at the dockside or airport is normally sub-contracted to specialist freight forwarding agencies. All this may impinge on engineers through the seemingly pedantic but absolutely necessary tasks of ensuring that all product labelling, packaging and documentation follows strict standards of conformance in order to avoid costly delays in the shipping process.

Engineering support to Distribution

Distribution is likely to be provided with the support of engineering services in the following manner:

1. Product Engineering can be involved by specifying the storage conditions (e.g. temperature and humidity controls) and the shelf life of a product.
2. Quality Control will be involved in tracking and reporting on rejects returns, warehouse or transportation damaged goods and out-of-shelf-life items.
3. Manufacturing engineering may design, specify and provide racking layouts, mechanical handling systems and packaging processes.
4. Maintenance engineering will maintain vehicles and mechanical handling or packaging equipment.
5. Facilities engineering will provide the warehouse with heating, ventilation, lighting and other general site services.

Summary

Distribution is an arm of the marketing/sales role charged with getting the product to the customer. Organisationally it is often part of one of those functions. Distribution can be a large element in businesses with international markets spread around the world or a matter of local delivery for a sole trader. In the former case, it can be a very complex and costly operation whose costs must be absorbed in the product price. Such an operation will involve engineering support in a variety of ways as outlined above.

4.5 Purchasing function

All businesses have a purchasing function. It may be very large, buying materials for resale in a retailing business or raw materials and components for conversion in manufacturing businesses. Alternatively, the purchasing function can be relatively small and confined, for example, to the purchase of stationery for a small consulting office. We will concentrate on the purchase of items in a company manufacturing engineering goods.

Figure 4.9 shows the organisation chart for such a purchasing function. Where the amount of purchasing warrants the employment of a number of buyers, they are usually organised into specialist sections. This allows each buyer to build up detailed knowledge of the potential suppliers and their prices and gives continuity of expertise and contact when problems arise with materials received. Purchasing is often linked with inventory control to form an overall materials management function. Figure 4.10 shows such an organisation.

Functions of Purchasing

The functions of Purchasing comprises:

1. Receiving notification of quantity requirements.
2. Sending of technical specification and delivery requirements to potential suppliers for the production of quotations or contract tenders.

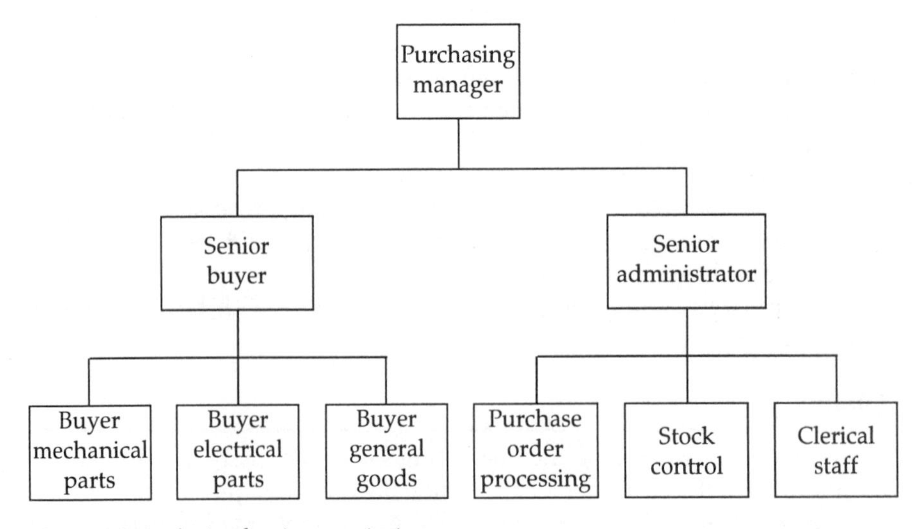

Figure 4.9 Purchasing function organisation

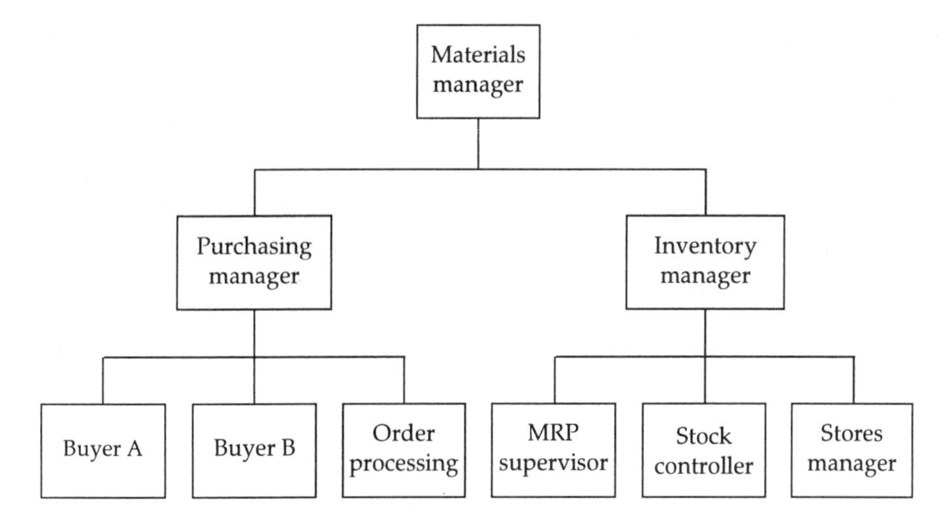

Figure 4.10 Combined purchasing and inventory control structure

3. The comparison of quotations and tenders. Examination of suppliers' contract conditions.
4. Selection of suppliers with agreement of the internal 'customer'.
5. Placement of order.
6. Communication on any queries or problems arising during completion of order by the supplier.
7. Receiving and noting confirmation of delivery from goods-in and acceptance by goods-in inspection.
8. Confirmation to accounts that the supplier's invoice can be paid (or not in the case of delays, rejects, etc.).
9. Maintenance of suppliers and order records.

Factors that a buyer will consider in selecting a supplier will include the following:

1. Quality guarantees
2. Quantity deliveries.
3. Price and discount structures, payment requirements.
4. After-sales service from the supplier.
5. Agreement and handling of any rejects.
6. Quality of packaging.
7. Proximity and ease of transportation and access.
8. Previous record as a supplier to the company.
9. Reputation as a supplier.
10. Pre-eminence in the field.
11. Financial stability of the supplier.
12. Check on financial soundness and debt record of suppliers via agencies that specialise in analysing the financial strength of companies.

Types of order

Orders to suppliers will take one of the three common forms:

1. Contract agreements for the regular supply of production materials or consistently used items like electricity and printed forms. Production and packaging materials requirements are nowadays usually calculated in a computer-based MRP (materials requirements planning) software program. The program is based on a master production schedule (MPS), itself derived from the sales forecast. The program will issue purchase quantity suggestions that are normally vetted by the buyer in order to gain the best commercial conditions as outlined above. Such programs are normally run on a regular weekly or monthly basis. Usually, an overall quantity and delivery schedule will be agreed for the life of the contract. However, the actual delivery from the supplier may be varied or 'called off' to the customer's requirements within certain agreed limits on quantity and frequency.
2. Separate but regular orders for items that have a steady usage over time, but whose consumption cannot be geared directly to the production schedule. These orders are often triggered by re-order levels being reached in the users' stores stock control system. Purchase requisitions (i.e. requests) are then sent to Purchasing department by the user store. This type of order may also be automatically produced by a computer-based stock control system. Orders may be placed for suppliers' standard delivery quantities so as to pay an economic price. In engineering functions this type of order may be for cutting tool replacements against the tooling stock re-order levels.
3. Separate orders for random supplies that are required occasionally. For example, these could be for a fixture, machine, or advertising literature. Again, the department or user will fill in and send an internal purchase requisition to the Purchasing department for the item.

In the case of special items like machines or new materials, engineers will, as a result of engineering investigations, make *de facto* decisions on who the supplier should be. It is wise, however, for every order to have the input of a buyer to ensure that no disadvantageous supplier's contract conditions are being accepted by the engineer. This is a real problem with potentially serious financial results if, perhaps young engineers without commercial experience are allowed to order materials, components or equipment directly, thereby committing the company to contractual conditions.

Contracts engineering

A significant proportion of engineering work is carried out against agreed contracts for one-off projects rather than in-house engineering for batch or volume manufacturing in a factory. Civil and mechanical engineering projects such as the building of bridges or factories are normally agreed and managed as a single contract with a main contractor responsible to the customer and/or the customer's architect. The suppliers of components such as the steel girders for a bridge or the boilers for a factory become involved as sub-contractors responsible to the main contractor. Project management techniques are widely used in the management of design, construction, commissioning and cost controls in this type of work.

The agreement and managing of the main contract, sub-contracts and purchasing are specialist areas requiring much practical experience and commercial acumen. This work is normally done by senior staff such as project directors or project managers. A junior engineer is most likely to first meet this type of work as an assistant to a project engineer working on one aspect of the contract. Contract documents are usually complicated in their attempt to provide financial and operational protection and guarantees to both sides under any foreseeable circumstance. Great care is taken therefore in the detail drafting of legal clauses and the schedules of specifications, work and materials to be used. The Institution of Mechanical Engineers has published standard forms of various contract agreements that potential suppliers and customers can use as a basis for their particular contract.

Engineering Support for Purchasing

Purchasing receives and distributes engineering information in several ways. Product Engineering sends part drawings and material specifications to Purchasing for the purpose of onward transmission to potential suppliers for quotations and contracted suppliers for the manufacture of the part. Manufacturing Engineering sends jig and tool drawings for the onward transmission to sub-contracted toolmakers. Quality Engineering will do the same for gauges and other quality assurance equipment. Facilities engineering will supply drawings and technical details of site services and maintenance equipment required.

Summary

Purchasing is a key function in any company because it provides the systems for the organised and controlled procurement of all the parts, materials and consumable items used by the business. A professional purchasing operation can ensure that the materials are purchased at the lowest possible cost related to the quality and quantities required, keep the flow of orders and materials in the right time sequences, maintain a base of knowledge of potential suppliers and ensure that only advantageous contract conditions are accepted by the company. If purchasing was left to the individual users of the supplies, it is certain that such benefits would be lost in a mass of competing and unprofessional purchasing practices.

4.6 Exercise 4

1. Describe three of the main activities of the Marketing function.

2. Outline the two functional levels at which Marketing can sit within an organisation.

3. What is the essential difference between the Marketing and Sales functions?

4. Of the possible development programmes of products and market share, which entails:
   (a) the least risk – and why?
   (b) the most risk – and why?

5. What is Marketing's relationship with Research and Development?

6. Name four key objectives of the sales function.

7. What are the alternative ways in which a sales function can be subdivided in organising its contact with the customers? What is the advantage of organisation by product type?

8. What is the major purpose of a sales forecast?

9. How important is a sales forecast to these other functions – Finance, Production and Purchasing?
   (a) Very important
   (b) Moderately important
   (c) Not important

10. From the sales forecasts in Table 4.2, work out the seasonal variation in total units and revenue for quarter three and quarter four, taking the average of the first two quarters as a base.

11. Name six product types that may be subject to seasonal variation in demand.

12. How can the Product Engineering and Quality Assurance functions support the Sales function?

13. Which two functions might Distribution most logically be part of in a company designing and manufacturing consumer goods?

14. Name product types that are likely to entail a large distribution function.

15. How might manufacturing engineering support the Distribution function?

16. What input would Quality Control give to Distribution?

17. Outline at least four key functions of the Purchasing department.

18. Describe the nature of two types of purchase agreement commonly used.

19. Which type of computer program is used to inform Purchasing of the production materials that should be ordered?

20. List four factors that a buyer should consider when deciding where to place an order.

21. Why is it important for a buyer to be involved in all purchases and in particular the purchase of technical goods such as capital equipment?

5

The financial function

5.1 Introduction

A majority of businesses are financially structured as limited companies and are entities separate in law from their shareholders, directors, managers and employees. All these persons may change completely, but this would still leave the company existing as a body in its own right with continuing legal and financial obligations. This is because the assets and monies in a company belong to the company and not to any individual. It is essential, therefore, that good financial control is exercised not only for the immediate survival and profitability of the business, but also to present an accurate picture of its financial assets and liablilities to its shareholders, suppliers, customers, and employees.

Detailed financial information is also required to deal with government tax gathering agencies such as the Inland Revenue, Department of Health and Social Security and Customs and Excise. Changes in government legislation on tax and other financial issues can cause a great deal of work for the financial function.

All these considerations give a significant role to the financial control of businesses. Such a task can only be successfully undertaken by applying a series of legally approved and commonly understood accountancy practices to the operations of the business. This is usually achieved by appointing qualified accountants (aided by clerical staff), who are either employees of the company or a professional accountancy practice. Small companies often do not employ a qualified accountant but rely on the services of an accounting practice. Medium-sized and large companies will normally employ qualified accountants but will also use an accountancy practice. In every case, however, the law requires that the company's financial books and position are audited by an independent body of accountants who have to certify that the position stated is correct.

Accountants are concerned with two main activities within the finance function, namely, financial accounting and management accounting. The Financial Accountant generally oversees the day-to-day financial management of the business and preparation of statutory accounts. This includes controlling the flow of funds (cash flows), meeting tax liabilities, arranging external financing, investing for the organisation and producing the annual accounts which include the balance sheet and profit and loss account. Section 5.2 contains a brief description of these but it is unlikely that engineers below senior management level will have much involvement in the considerations of financial accounting.

The Management Accountant deals with the financial control of the day-to-day operations of departments and activities within the organisation. This is the aspect of accounting that non-accounting managers and engineers will be involved in because it provides the mechanisms and data needed to forecast, record and control operational costs and income. As such, it is the area of financial management that engineers are most likely to meet in their work and sections 5.3 onwards are concerned with management accounting as it relates to the engineering functions. Figure 5.1 shows a possible organisation chart for the financial function of a medium-sized company.

5.2 Financial accounting

Flow of funds

The balance sheet and profit and loss accounts are part of an annual statement and summary of the financial situation. They are the result of the flow of funds during the year and it is this that must be controlled on a continuous basis throughout the year to ensure financial success.

Figure 5.2 shows a typical flow of funds for a manufacturing company. These are known as circulating assets.

- *Creditors* – lenders such as the banks, suppliers of materials and services and employees' due salaries.
- *Stock* of raw materials – this is the value of the stock held in the form of raw materials.
- *Work-in-progress (WIP)* – the value of materials in the process of conversion within the factory.
- *Finished goods* – saleable product that is held in store or is in shipment but still owned by the company.
- *Debtors* – predominately the money that customers owe the company for the supply of the product.
- *Cash flow (profit)* – net receipts available for expenditure after payment of the creditors.

Balance sheet

The balance sheet provides an overview of the financial position with a snapshot showing the assets and liabilities at a particular time. All companies' accounts show a balance sheet at the accounting reference date

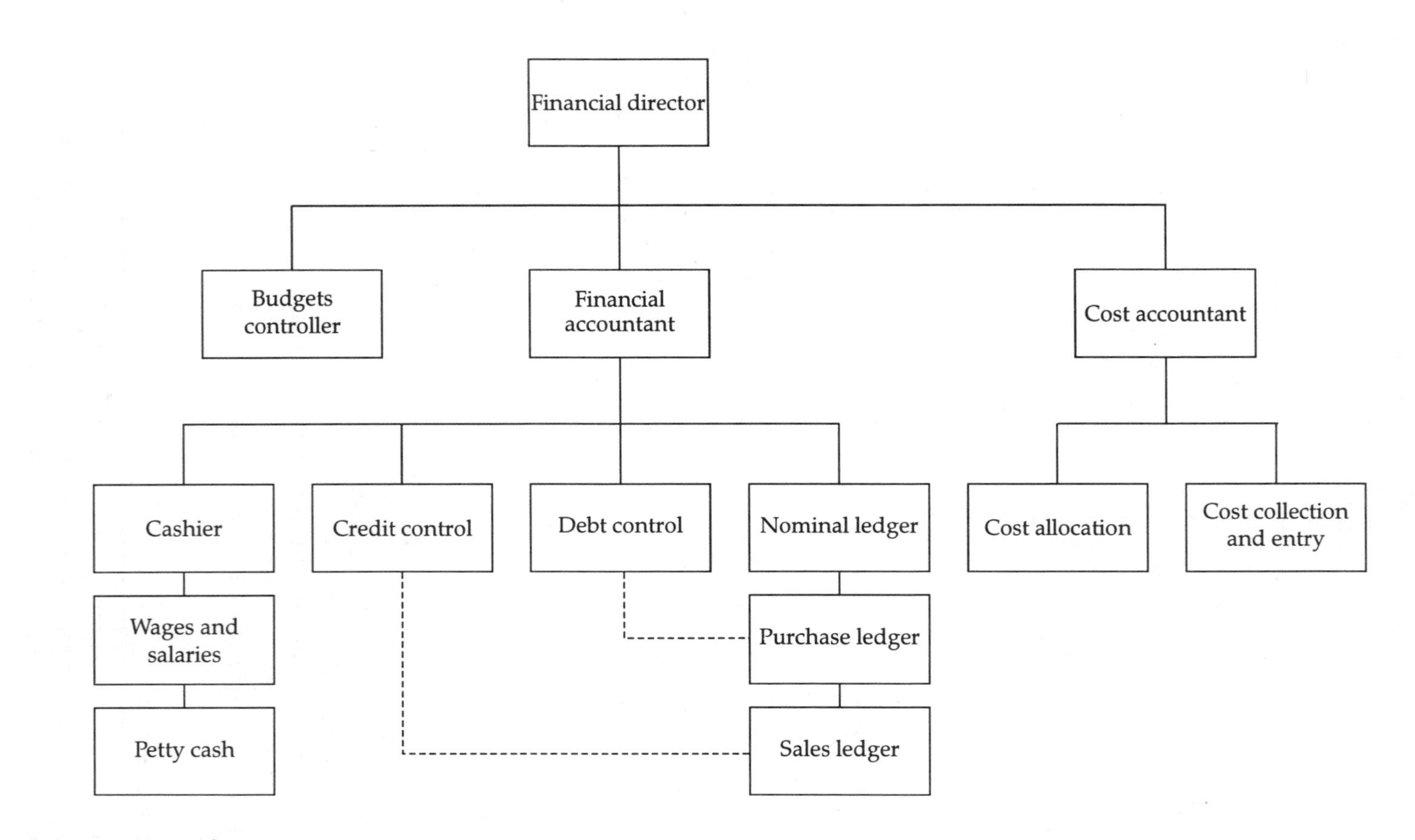

Figure 5.1 Financial function organisation structure

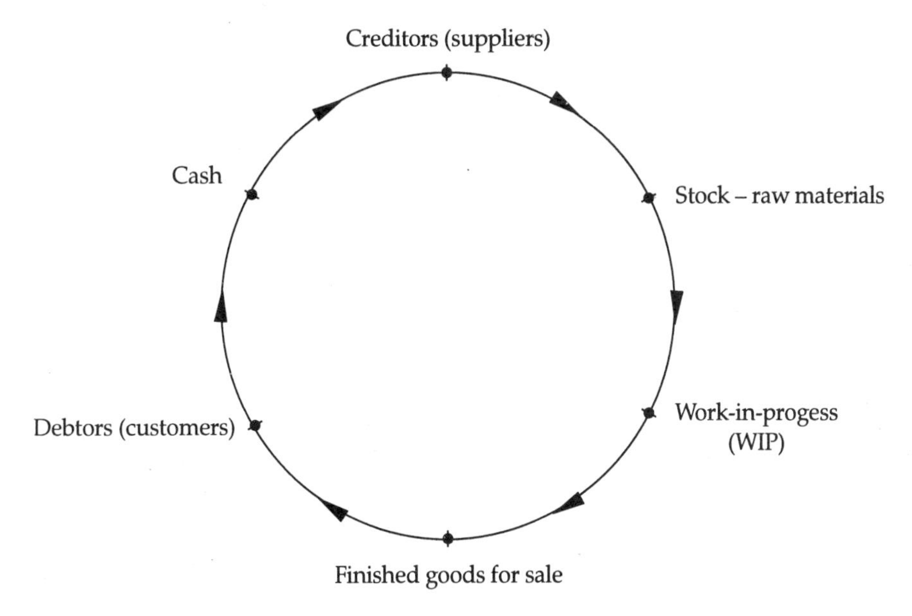

Figure 5.2 Circulation of assets

(known as the reporting date) reflecting the changing position of assets and liabilities at the end of the period covered by the profit and loss account. In most organisations this is a 12-month period starting at the date of incorporation (set-up) of the company or the commencement of trading, not the calendar year.

In the UK the requirement to produce a balance sheet was introduced by law in order to give investors in limited liability companies evidence of the financial status of the company. Successive Acts of Parliament have increased and refined the amount of information that balance sheets and profit and loss accounts must contain, to the point where they can be large and complex documents.

The simple balance sheet in Figure 5.3 illustrates the main items included. These are defined as either assets or liabilities. The totals of these must balance (hence the name), if the financial position of the organisation is sound and properly controlled. Thus the books balance with a total of £60 000 shown on both sides of the balance sheet in the example shown.

Profit and loss account

The profit and loss account summarises the trading position and operating income and expenditure during the financial period under review. The final balance of profit or loss is then included in the balance sheet as a component of the overall financial situation. Table 5.1 shows the P&L account for A Small Company Ltd. The account shown is a simplified version but it illustrates the essential components. The format is to start with the business

Liabilities	£	Assets	£
Subscribed Capital and Reserves		*Fixed Assets*	
Ordinary shares	30000	Properties	20000
Reserves	12000	Plant and Machinery	15000
Profit (from profit and loss a/c)	8000	Vehicles	10000
Shareholders funds	50000	Total fixed assets	45000
Current liabilities		*Current assets*	
Creditors	7000	Stock	5000
Bank overdraft	3000	Debtors	10000
Total	10000	Total	15000
Final total	60000		60000

Figure 5.3 Balance sheet for A Small Company Limited

income from sales in the year and subtract the various expenditures to arrive at a net profit after tax. Dividends would then be paid from the profit and the remainder retained as reserve. The percentage figures are a useful indication of the scale of the expenditures and profit as a percentage of the sales value.

Sometimes financial items such as interest earned from investments and paid as charges may be shown separately as will depreciation on capital. This is done to isolate these financial charges from the mainstream of the business activity and present a trading profit figure. This gives a more accurate measurement of the profitability of the main business of the company. It is also usual to show last year's figures for comparative analysis. Table 5.2 illustrates such a P&L account. The data in Table 5.2 seem to give us a

Table 5.1 Profit and loss account of A Small Company Ltd: year ending 31 December 1996

	£	£	% Sales
Sales		100000	100.0
less Expenditure			
Materials	35000		35.0
Wages and Salaries	45000		45.0
Overheads	9800		9.8
Total		89800	89.8
Gross profit		10200	10.2
less Tax		2200	2.2
Profit after tax		8000	8.0

reasonable amount of information about the levels of expenditure and the changes in them between 1995 and 1996 which are shown in Table 5.3.

Unfortunately, the figures only show the *results* of one year's trading. They do not tell us *why* the changes happened. The changes could have been due to price and wage rises for the same level of output. Alternatively, output may have risen and efficiency improved to give the greater rise in profit (96 per cent) than costs (5.9 per cent).

To take an analogy from football, it is like having the final score of a 3–2 win, without knowing how the teams or individuals played throughout the match. Did the strikers score well? Did our goalie let in two soft goals or not? Did the manager's second-half tactics work better than the first half? In short, what lessons have we learnt for our next match?

Table 5.2 Profit and loss account: 1996 and 1995 comparison

		1996		1995
Sales		100000		90000
less expenditure				
Materials	35000		33000	
Wages and salaries	45000		42000	
Overheads	9800		9800	
Total		89800		84800
Gross trading profit		10200		5200
plus interest received	800		800	
less interest paid	500		400	
less depreciation	2000		2000	
		1700		1600
Gross profit		8500		3600
less tax		2000		1000
Profit after tax		5500		2600
Dividends		1500		600
Retained profit		4000		2000

Table 5.3 Changes for 1995

	£	%
Sales+	+10000	+11.1
Materials	+2000	6.0
Wages and salaries	+3000	7.1
Overheads	0	—
Total	+5000	5.9
Gross trading profit	+5000	96.1

A final score can only be an indicator and not a detailed account of what happened during the 90-minute match or one year of business operations. Thus, in order to arrange the business more efficiently and profitably, more detailed financial information is needed than can be seen from the balance sheet or profit and loss account. This is where management accounting and various decision-making techniques come to play important roles which we will now explore.

5.3 Management accounting

This is concerned primarily with accounting for and controlling the internal costs of the business and reporting the financial situation to management. The person in charge of management accounting is usually called the management accountant or cost accountant. Management accounting provides the financial data regarding the detailed operational activities of the business by maintaining a costing system that does the following:

- forecasts costs through budgets;
- records actual costs incurred;
- reports on costs to managers;
- provides the basic data for the preparation of the profit and loss account.

The financial data obtained from a costing system support and compliment the quantitative and qualitative data available from operations. In other words, costs provide a third piece of information in addition to physical quantities and performance measurements. Typical sources of costs data are:

- Material costs
  - Suppliers' invoices
  - Internal material requisitions
  - Stores issues reports
  - Scrap reports
- Labour costs
  - Payroll information (employees and wage/salary rates)
  - Time bookings on jobs
  - Overtime bookings
  - Incentive scheme awards
  - Sickness bookings
  - Pension details
  - National Insurance details
- Overhead costs
  - Will include any of the above types of material and labour cost bookings that contribute to the costs of overheads.
  - Suppliers' invoices for rent, electricity, gas, water, etc. plus accountancy and other external professional fees, stationery, travelling expenses, etc.

How costs are budgeted, calculated and controlled is explored in greater detail in Chapters 7 and 8 (which corresponds to Unit Element 1.3). For the remainder of this chapter, we will concentrate on various financially-based management decision-making techniques.

5.4 Financial decision-making

Money is the life blood of business and is spent in two basic ways.

1. Capital expenditure.
2. Operating expenditure.

We shall now examine how decisions are taken with regard to both of these.

Capital expenditure decision-making

Capital expenditure is the payment for investments in fixed assets such as buildings and production equipment. Such expenditure is a series of one-off investments, each being justified on its own merits *vis à vis* operational necessity or cost savings. Once spent, the money is not wholly recoverable in the short term through sales income and has to be 'recovered' gradually over a number of years via the benefits that justified the investment.

Capital investment is needed for a variety of purposes and many businesses group their investments into categories based upon the main purpose of the expenditure, as shown in Table 5.4.

The classification is an important element because each may be subject to a different policy outlook from senior management. For example, replacement may not be approved if it is known that the resource will not be needed after next year. An item classified as a necessity is likely to be approved for legal and policy considerations rather than for any cost benefit. In any case there may be none directly obtainable.

The approval of a capital expenditure involves six stages.

1. Collection of data on the existing situation.
2. Collection of data on the possible options for action.
3. Calculation of the financial consequences of each option.
4. Selection of the best alternative.
5. Writing a request for the capital expenditure to be made.
6. Submission and approval of the request.

Items 1–5 inclusive are usually the responsibility of one person, possibly the operating manager of the area to have the investment, a designated subordinate or often, with capital expenditure on manufacturing equipment, the appropriate manufacturing engineer.

The review is carried out by the circulation of the written request to the people listed below.

Table 5.4 Investment categories

Category	Main purpose
New products	The introduction of new products and the additional facilities needed to design, produce and sell them.
Expansion	Additional facilities to undertake more design, manufacture or sales of the existing products.
Cost reduction	Investment aimed at gaining cost savings over existing methods and equipment used.
Replacement	Investment needed to replace facilities that are wearing out or no longer economic.
Necessity	Unavoidable investments to meet health and safety legislation, personnel requirements, environmental legislation and social obligations to the community.

- The operating manager responsible for the area in which the investment is to be made.
- Senior functional management for the same area.
- The Financial Director or equivalent.
- The MD or General Manager.

We will now examine the stages in turn using the following example. The marketing department of a company making pumps indicates that sales could increase from the existing 10 000 to 12 000 p.a. However, it appears that the factory will reach the limit of its present productive capacity at 11 000 units per year. The company must decide whether to:

1. Restrict the future sales level to 11 000.
2. Purchase more production machinery to produce the extra 1000 p.a.
3. Sub-contract the manufacture of the extra 1000 to another company.

The company will need to financially evaluate all three options. Option 2 will involve capital investment and an investigation would proceed as follows.

Stage 1 Collection of data on the existing situation

Is the current capacity REALLY limited to 11 000 p.a. or can production do any of the following?:

(a) increase overtime;
(b) introduce shift work;
(c) improve the output of existing equipment (with the help of manufacturing engineering).

Stage 2 Collection of data on possible options for action

If the answers to stage 1 are no, or only with unacceptable risks of not reaching the 11 000 p.a. target, then additional capacity must be provided. If so, the manufacturing engineer will need to investigate:

(a) what equipment will be required technically;
(b) what its maximum output will be;
(c) how much it will cost to buy and install;
(d) what labour and support services will be needed;
(e) how much it will cost to run including the labour and support costs;
(f) if there is sufficient space available for
  (i) the additional equipment for production and packaging;
  (ii) the additional materials' storage and handling.
(g) when the installation could be operational.

Stage 3 Calculation of the financial consequences of the options

Because of the long-term importance of investment decisions, several methods of calculating the financial consequences of such projects over a number of years have been developed. These are examined in the next section, on p. 98.

Stage 4 Selection of the best option

It is likely within any investigation that several possible solutions will present themselves. It will be necessary therefore to carry out stages 2 and 3 on a number of possible solutions in order to be able to compare them and produce a final recommendation for investment in the request for capital expenditure.

Stage 5 Writing of a request for capital expenditure

Most companies that regularly make capital investments have a formal system of capital expenditure approvals. This involves the writing of a formal document asking for permission to adopt the option recommended and spend the money. The document is often called a 'Request for Capital Expenditure', 'Authority for Capital Expenditure' or similar name. Normally it is written by the engineer or manager responsible for the bulk of the investigative work and it is then circulated to various levels of management for approval of the proposal. A typical request should include:

1. A first page summary which shows
   (a) reasons for the request;
   (b) summary of the proposal;
   (c) summary of the benefits.
   Note: In business, all reports and submissions for decisions to management should start with a summary so that busy senior managers can grasp the critical points without reading the whole submission. They can then read any of the detailed body of the document if they need clarification on some point. If a summary is not included, the most likely result is a considerable delay at each approval stage.
2. A fuller description of the proposal with supporting data on:
   (a) current situation and problem;
   (b) detailed description of the proposal;
   (c) detailed description of the benefits.

3. Financial calculations including any Discounted Cash Flow (DCF) figures.
4. Supporting appendices and relevant reports.

Stage 6 Submission and approval of the request

The request is normally circulated to appropriate levels of management via a documentation system controlled by Accounts or quite often Engineering Administration. The originator and his/her manager plus the operating manager who will use the investment will all sign the request. It will then be circulated to ascending levels of manager for approval. The signatories to the document are normally decided by the level of expenditure involved. Each level of manager is usually assigned a maximum for which their signature is all that is necessary. Any amounts above this must be approved by the next higher level of signatory. For example, in a medium to large company the signing regime might be as shown in Table 5.5. Requests are only approved when all the signatures up to and including the most senior required have been obtained.

Table 5.5 The signing regime

Position	Approval limit
Department managers	£2000
Function managers	£5000
Function Director	£10000
Financial Director	all requests
Managing Director (on own)	£100000
Board Decision	all requests above £100000

On a large project with a significant capital investment, the investigation and writing of the request can take up to one year and the approval routine a further three months. It is essential therefore that the likelihood of such projects is anticipated and an approximate sum forecast in the capital expenditure budget for the likely year(s) of expenditure. Capital budgeting is discussed in Chapter 8.

Investment appraisal

The investment appraisal methods commonly used to evaluate capital expenditure proposals are:

- Payback method
- Net Present Value (NPV)
- Internal Rate of Return (IRR)

Payback method

The payback method only measures the degree of risk in terms of time taken to recover the cost. It does not indicate profitability. An example is given in Table 5.6.

Table 5.6 Example of the payback method

		Project A	Project B
Capital cost		200000	200000
	Benefits Year 1	100000	60000
	Year 2	100000	70000
	Year 3	80000	70000
	Year 4	60000	110000
	Year 5	40000	120000
Totals in 5 years		380000	440000
Payback period		2 years	3 years

Although project B has the longer payback period of three years, it is clearly more profitable in the long run. Thus, a decision taken via the payback period method could be in error if long-term benefits are not considered.

The concept of net present value (NPV) and discounted cash flow (DCF)

When planning investments from which returns will take more than one year to come in, an important factor that must be taken into consideration is the gradual devaluation of money over time. £1.00 received in a year's time is worth less in purchasing power than £1.00 received today. In other words the real value of the money will have declined to a percentage of the present value.

The present value of one unit of money at inflation rates of 4 per cent p.a. (near the current rate) and 10 per cent (applicable in the 1980s) is shown in Table 5.7.

The formulae for calculating the annual present values is

$$PV = \frac{C}{(1 + r)^n}$$

Where PV = present value; C = cash flow; r = rate of interest; n = number of years.

The formula for the summation of n number of years of project life is:

$$PV = \frac{C_1}{(1 + r)} \quad \frac{C_2}{(1 + r)^2} \quad \frac{C_3}{(1 + r)^3} \quad \cdots \quad \frac{C_n}{(1 + r)^n}$$

A table of the present value of one money unit for a range of discount rates and years hence is given in Appendix 1, on p.165.

It can be seen that the cumulative disadvantages are considerable because the loss of value compounds with the increase in time. Thus, the reduction in value of money can have a significant effect on large long-term investments. For example, if cost savings of £1 000 0000 a year for 5 years is forecast for a project, then at 4 per cent inflation, instead of having £5 000 000 at present value after 5 years, the income will be £4 452 000. The face value will still be

Table 5.7 Present volume of money at inflation rates of 4 and 10%

Time period	4%	10%
Now	1.000	1.000
After 1 year	0.962	0.909
After 2 years	0.925	0.826
After 3 years	0.889	0.751
After 4 years	0.855	0.683
After 5 years	0.822	0.621

£5 000 000 but the real value at today's levels (i.e. present value) will be £5 000 000 – £4 452 000 = a reduction of £548 000 or 11 per cent.

It is important, therefore, for any long-term or large investments, that the reduction in money values is allowed for in the calculation of expenditures and returns. The process of doing this is called Discounted Cash Flow (DCF) and two methods are in common use.

1. The Net Present Value (NPV) method. This method calculates the future net value of a project expenditure and returns by using the discount factors demonstrated above. Table 5.8 shows the financial comparison between two projects. Project 2 clearly has a better return with a NPV of £1892.
2. Internal Rate of Return (IRR) also known as the yield method. This method calculates the rate of interest needed for the present value of the returns to be the same as the present value of the investment needed, i.e. that the net present value be zero. This is the point at which the project breaks even. Taking the figures from the NPV calculations in Table 5.8, these give us a result for one discount rate, i.e. 4 per cent. We need a second calculation at a different rate of discount in order to plot a graph of the IRR.

Thus Table 5.9 shows the same projects with a discount rate of 10 per cent applied. From these two sets of figures we are able to plot a graph of the IRR for each project as shown in Figure 5.4. The graph shows the IRR for

Table 5.8 Net present value method of comparing projects using 4% discount rate

Year		Rate @ 4% Discount	Project 1 £	Project 1 PV £	Project 2 £	Project 2 PV £
0	Investment	1.000	–10000	–10000	–12000	–12000
1	Savings	0.962	3000	2886	3500	3367
2	Savings	0.925	3000	2775	3500	3238
3	Savings	0.889	2500	2223	3000	2667
4	Savings	0.855	2000	1710	3000	2565
5	Savings	0.822	2000	1644	2500	2055
	Net total and present value		2500	1238	3500	1892

Table 5.9 Net present value method of comparing projects using 10% discount rate

Year		Rate @ 10% Discount	Project 1 £	Project 1 PV £	Project 2 3	Project 2 PV £
0	Investment	1.000	−10000	−10000	−12000	−12000
1	Savings	0.909	3000	2727	3500	3182
2	Savings	0.826	3000	2478	3500	2891
3	Savings	0.751	2500	1878	3000	2253
4	Savings	0.683	2000	1366	3000	2049
5	Savings	0.621	2000	1242	2500	1553
	Net total and present value		2500	−310	3500	−73

project 1 at a rate of 8.8 per cent and project 2 at 9.8 per cent. This illustrates again that project 2 is the best investment.

If one calculated point is positive and one negative, then the plot line will automatically cross the discount rate axis at the point of the IRR. If both plot points are above or below zero, the discount rate line may be extended to cross it. One of the advantages of this method is that any number of competing projects can be prioritised by the IRR that they give. Most companies using this method will set a minimum IRR that a project must show before it would be acceptable. Usually the IRR is automatically calculated in computer programs rather than by drawing a graph, but the graph has been used here as a better way of illustrating the concept.

In commenting on the use of capital investment decisions and resultant return percentages, some of the literature suggests that these are used to decide between the options of spending the money on the proposal versus leaving the money in the building society earning X per cent interest p.a. In reality, most companies do not have enough capital to do all they want and the purpose of capital expenditure calculations is to decide which of the competing internal projects warrants spending the available capital.

Depreciation

Depreciation is an important consideration in capital investment. The term is used to describe the decreasing value of assets over time. In particular, plant and machinery will wear out and also become technically obsolete, thereby needing replacement after a number of years.

Given that the total value of a company's assets may be £ millions, the figure for total depreciation can be very large. The depreciation figure is entered into the profit and loss account to show the cost and into the balance sheet to show the reduction in the value of the assets. A machine bought for £50 000 today will only be worth say £5 000 at resale after a few years of wear and tear. The company accounts must make allowance for this loss in company value through the depreciation of its assets.

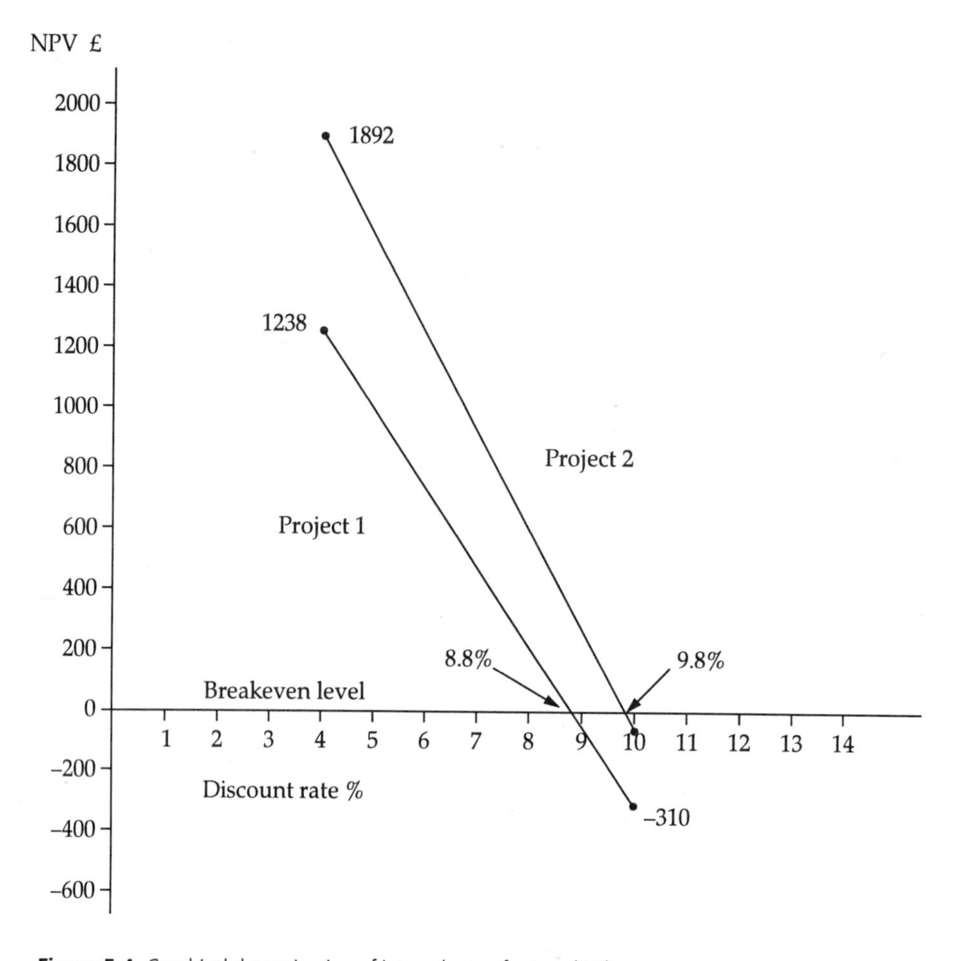

Figure 5.4 Graphical determination of internal rate of return (IRR)

Most businesses set rules for the depreciation rate per annum. A commonly used rate is 10 per cent (i.e. assuming a 10-year life for the asset). For equipment such as computers which become obsolete more rapidly, a 20 per cent (i.e. 5-year life) figure is often used. The Inland Revenue recognises the cost to business of asset depreciation by granting a tax-free capital allowance of 25 per cent of the asset cost per annum. This does not mean that the cost is covered by tax allowances in four years because the allowance is calculated on a reducing balance basis.

We will now examine three methods of calculating depreciation which are:

- Straight line method;
- Reducing balance method;
- Unit cost method.

Straight line method

This is based on the calculation of an equal amount of depreciation in each year.

Example

Cost of asset	£10000
Scrap value	£1000
Depreciation	£9000

Estimated life = 10 years

Annual depreciation = £9000/10 = £900 = 9% of £10 000 cost

The term straight line depreciation comes from the shape of the graph that this method produces. Figure 5.5 shows an example.

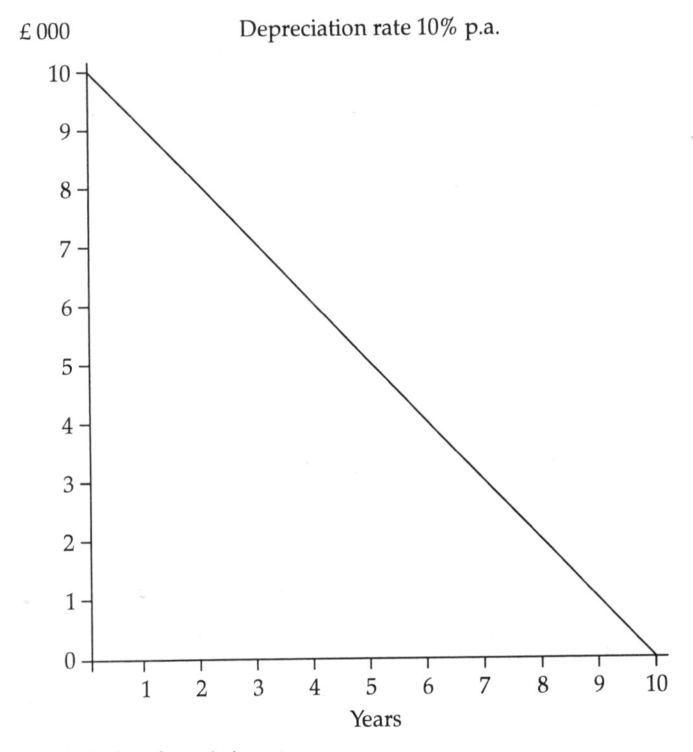

Figure 5.5 Straight line depreciation

Reducing balance method

In this case, the depreciation charge is taken as a fixed percentage of the asset value at the beginning of the year which is then reduced by the calculated percentage. This leaves a lower start-of-year figure for the next period and the percentage is then applied to this. In other words, the depreciation percentage is applied to the reducing asset value each year.

Example

Cost = £10 000
Depreciation rate = 10%

Year	Start value (£)	Depreciation at 10% p.a. (£)	Residual value (£)
1	10000	1000	9000
2	9000	900	8100
3	8100	810	7290
10	3874	387	3487

and so on

It can be seen that when using the same annual percentage as in straight line depreciation (10 per cent in this case), the reducing balance method produces a progressively smaller actual depreciation figure in £ and, therefore, takes much longer than the straight line method to reach the expected scrap value. Consequently, a rate of 25 per cent is commonly used for reducing balance calculations:

Example

Cost = £10 000
Depreciation rate = 25%

Year	Start value (£)	Depreciation at 25% p.a. (£)	Residual value (£)
1	10000	2500	7500
2	7500	1875	5625
3	5625	1406	4219
10	780	217	563

and so on

It can be seen from the curve on the graph in Figure 5.6 that the residual balance can never reach zero.

Unit cost method

In this case the depreciation is spread over the number of units the investment is expected to produce in its lifetime.

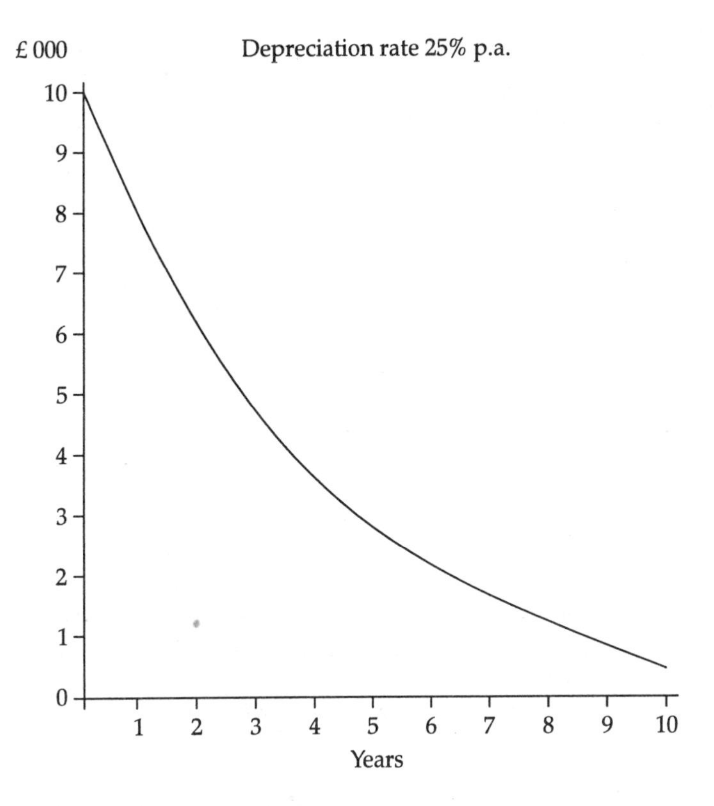

Figure 5.6 Reducing balance depreciation

Example

Cost	=	£10 000
Scrap value	=	£1 000
Depreciation	=	£9 000

Estimated output in life of asset = 900 000 units

$$\text{Depreciation charge} = \frac{£9\,000}{900\,000} = £0.01 \text{ per unit}$$

The problem with this method is that if the investment does not produce the full number of units, then the depreciation charge will be too low. Also, if the output volume varies significantly year to year, then the depreciation charge may be too low or too high and will need to be adjusted annually in order to be correct.

Key points

- **Capital expenditures** are essentially one-off costs that need to be kept separate from day-to-day operational expenditure for control, accounting and tax purposes.

- An efficient business will have a formal procedure for investigating, proposing and approving capital investment projects.
- Investment appraisal for projects lasting more than one year should be calculated using discounted cash flow methods in order to allow for the devaluation of money with time.
- Asset **depreciation** is an important factor that must be taken into consideration. Straight line and reducing balance methods apply an annual percentage figure.

5.5 Operating expenditure decision-making

Operating expenditure is that which is spent on carrying out the daily operations of the business functions in their tasks of designing, making, selling and supporting the product or service. Operating expenditures (i.e. costs) are classified into three categories depending on to what degree they vary with changes in the level of production output. The categories are:

- *Variable:* These change in direct proportion to the level of production output. The major elements of cost in this category are the direct labour and direct materials consumed in the output of products or services.
- *Semi-variable:* These vary disproportionally and often in stepped stages relevant to output. Indirect services and materials that support production can fall into this category. Maintenance costs are an example that may increase as certain thresholds of output volume are reached.
- *Fixed costs* for the financial period under consideration. For example, annual building rent and rates that must be paid regardless of the level of output. See Chapter 7, Cost Headings for costing in more detail.

Calculating profitability and the break-even point

One application of the cost information in operational decision-making is in the calculation of the profitability of particular products. This is calculated as follows:

Profit = sales revenue – total costs
Total costs = variable cost + semi-variable cost + fixed cost

Example

A company is able produce up to 500 units per year of a product that has a likely price of £10.25 each in the market place. The detailed costs are:

Variable	=	£ 4.25 per unit
Semi-variable	=	£ 100 for up to 100 units
	=	£ 300 for 101 to 300 units
	=	£ 400 for 301 to 500 units
Fixed	=	£1000 regardless of output level

We wish to know what the minimum quantity is that must be sold to break even and what the levels of profit will be for sales of 100, 200, 300, 400 and 500 per year.

From the data supplied we can construct Table 5.10. From the table we can see that a loss is made until over 200 units are sold. We can also draw the break-even chart shown in Figure 5.7. In this the three categories of cost are summed to give a total which can be plotted against sales revenue in order to show at what volume of output the receipts break even with the costs. Before the break-even point is reached a loss will be incurred, thereafter a profit will be made.

Table 5.11 shows the break-even point check calculations. The chart shows that the loss will be a minimum of the fixed costs, because the variable and semi-variable costs are not incurred until production commences. However, the fixed costs are already committed even if the product under consideration was not made. This leads to an accounting argument that says it is an error to add in the fixed costs when working out the profitability of additional or marginal production because those fixed costs are already incurred. If the fixed cost is excluded, then a contribution to profit of the revenue less variable and semi-variable costs is made as shown Table 5.12. This is a logical approach when considering the profitability of additional production/sales because it reflects the true additional cost to the business of the extra production. This approach is called marginal costing, of which more is said in Chapter 7.

Caution: Marginal costing cannot be used across the board to establish all basic prices because these must include an apportionment for fixed costs. Otherwise these will never be recovered from the customer and the business would make a loss of at least the value of the fixed costs.

Table 5.10 Profitability calculations

Sales level (units)	**100** **£**	**200** **£**	**300** **£**	**400** **£**	**500** **£**
Revenue @ £10.25 each	1025	2050	3075	4100	5125
Variable costs					
Cost per unit	4.25	4.25	4.25	4.25	4.25
Total variable cost	425	850	1275	1700	2125
Semi-variable cost	100	300	300	400	400
Fixed cost	1000	1000	1000	1000	1000
Total costs	1525	2150	2575	3100	3525
Profit or loss	–500	–100	500	1000	1600

Table 5.11 Break-even check calculations

Sales level (units)	216 £	217 £
Revenue @ £10.25 each	2214.00	2224.25
Variable costs		
Cost per unit	4.25	4.25
Total variable cost	918.00	922.25
Semi-variable cost	300.00	300.00
Fixed cost	1000.00	1000.00
Total costs	2218.00	2222.25
Profit or loss	–4.00	2.00

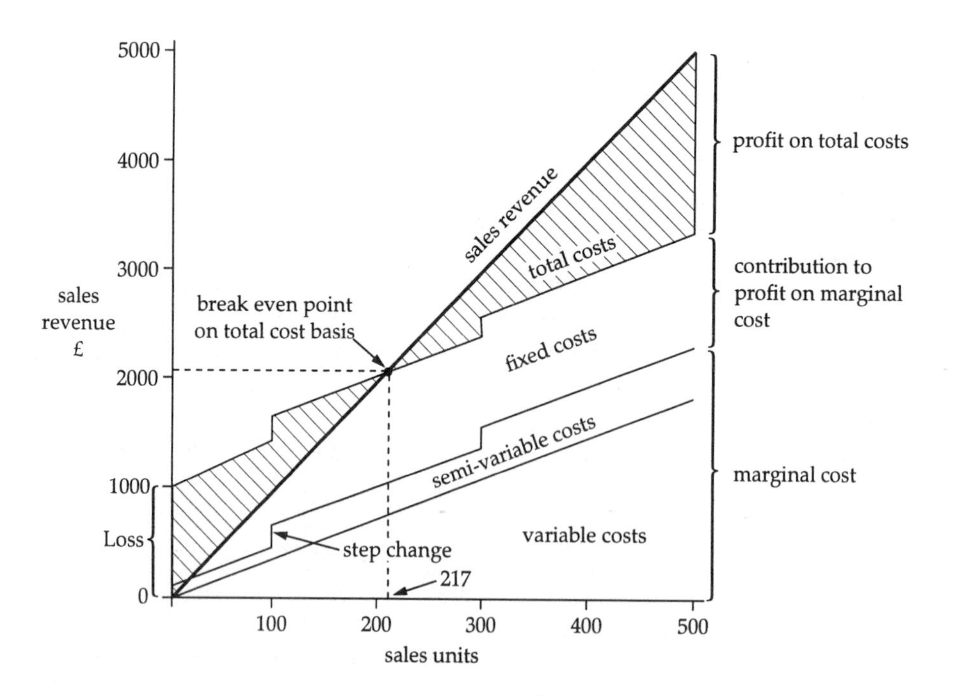

Figure 5.7 Break-even chart

Table 5.12 Profitability calculations: marginal contribution basis

Sales level (units)	**100** £	**200** £	**300** £	**400** £	**500** £
Revenue @ £10.25 each	1025	2050	3075	4100	5125
Variable costs					
Cost per unit	4.25	4.25	4.25	4.25	4.25
Total variable cost	425	850	1275	1700	2125
Semi-variable cost	100	300	300	400	400
Total costs	525	1150	1575	2100	2525
Contribution	500	900	1500	2000	2600

Make or buy decisions

This type of decision needs to be made when:

- a new product or part is being introduced;
- production capacity is limited;
- cost reductions are being sought.

Here a comparison between the cost of making the part and purchasing it must be made. At the simplest level, the comparison will consider the problem on a marginal basis:

Example

Cost of manufacture/unit		**Purchase costs/unit**	
Direct labour	6.40	Purchase price	17.30
Direct material	8.50	Purchasing overhead	1.50
Variable overhead	2.20		
Ex-factory cost	17.10	Purchase cost	18.80

Saving over purchase = £18.80 – £17.10 = £1.70

Using the marginal cost basis, the advantage of £1.70 per unit maintained regardless of the quantities involved.

If fixed overheads are added to the cost of the manufacture, then the decision would be reversed.

Ex-factory cost	£17.10
Fixed overhead	£2.50
Total cost	£19.60

On the basis of this calculation it would be cheaper to purchase the part. The point to be made here is that the business accounting practices on overhead apportionments can have an important effect on the decision. It is essential therefore that all make or buy decisions follow a common accounting policy on the inclusion or not of variable and fixed overheads in the computations of the manufacturing and purchasing costs.

Key points

- Operating decisions can be made on the basis of cost calculations involving variable, semi-variable and fixed costs.
- The inclusion or not of fixed costs will have a significant effect on the outcome.
- Marginal costing (which excludes fixed costs) can be used to calculate the additional cost of production or other activities using the assumption that the fixed costs are already incurred whether the additional production takes place or not

5.6 Exercise 5

1. Describe the two main branches of finance and their primary roles.
2. Draw a diagram illustrating the concept of circulating assets.
3. Which two financial statements are normally produced for every limited company?
4. Explain the purpose of a balance sheet. What must the mathematical result be? If a business only produces one edition of the balance sheet in a year, when will this be valid for?
5. Construct a balance sheet to include the following elements.

Share capital	£50 000
Profits for P & L	£10 000
Creditors	£9 500
Total current liabilities	£12 000
Total fixed assets	£48 000
Stock	£10 500

   Following the structure shown in Figure 5.3, insert your own figures for the elements that are missing to give a total balance of £100 000.
6. What does the profit and loss account do?
7. Give the three cost categories that are normally recorded in the costing system of a design and manufacturing company.
8. Explain the difference between capital expenditure and operational expenditure.

9. Describe the main purpose of three of the categories of capital expenditure that may be found in business.

10. In an investigation of capital expenditure on a new and faster production machine, which cost data should be sought to calculate the cost savings?

11. List the three methods used in investment appraisal. What are the advantages and disadvantages of the payback method?

12. A capital investment investigation into the purchase of a machine produces the following data: Capital cost £100 000 – to be spent in one lump sum. Cost savings forecast – £ 32 000 p.a.
    (a) Prepare a net present value using a 5 per cent for a 5-year period.
    (b) Establish the internal rate of return (IRR).

13. A company has a capacity problem and is considering purchasing a part in order to relieve demand on the machine shop. Given the following data, decide which course of action should be taken. Fixed costs can be excluded in this case.
    Direct labour cost £8.50 per unit.
    Direct material cost £3.00 per unit.
    Variable overhead 15% of direct labour.
    Purchase price £10.50.
    Purchase overhead 12% of purchase price.

14. The production and sale of a product carry this data:
    Output – up to 500 p.a.
    Sales value £10.00 each.
    Variable cost £5.00 each.
    Semi-variable cost £1.00 for 1 to 199 p.a., £2.00 for 200 to 1000 p.a.
    Fixed cost £500.
    Draw up a calculation table and a break-even chart to determine:
    (a) the break-even quantity of units over the total costs;
    (b) the profit on sales of 100, 200, 300, 400 and 500 units;
    (c) the contribution on sales of the same quantities;
    (d) the profit and the contribution of the break-even quantity.

6

Interfaces and information flows between functions

6.1 Introduction

So far, we have examined the functions most commonly found in businesses. We will now examine the interfaces and information flows between them. An important concept to understand as the basis for these relationships is that, internally, a business is a self-contained economy and within it each function, department, section and person has customers and suppliers of its own. We will analyse the relationships of each function in turn, following the logic of the product flow from R&D to after-sales service. It will also be useful to return to the organisation charts in Chapters 2 to 5 when studying a particular relationship. All business functions interface with two service functions, namely, personnel regarding the recruitment and conditions of employees and the finance function in regard to the creation and monitoring of their operating and capital budgets. In order to avoid repetition, these interfaces are not included in the following lists. The key interfaces between functions are illustrated in Figure 6.1.

6.2 Interfaces of engineering functions

Research and Development (R&D)

R&D will receive guidelines from senior management and marketing on the direction of its work. In well-organised companies this will be done through the medium of some sort of corporate product objective issued by a new product policy team comprising members of senior management, marketing and the R&D department.

R&D will feed back the results of preliminary research findings and feasibility studies to the 'new product policy' decision-makers. A decision can then be taken whether to proceed to fuller research in the likelihood that a new product launch will be made. R&D programmes need to be systematically controlled in this manner, otherwise substantial sums of money can be spent in an open-ended quest for knowledge that does not at some point transmit itself into saleable products.

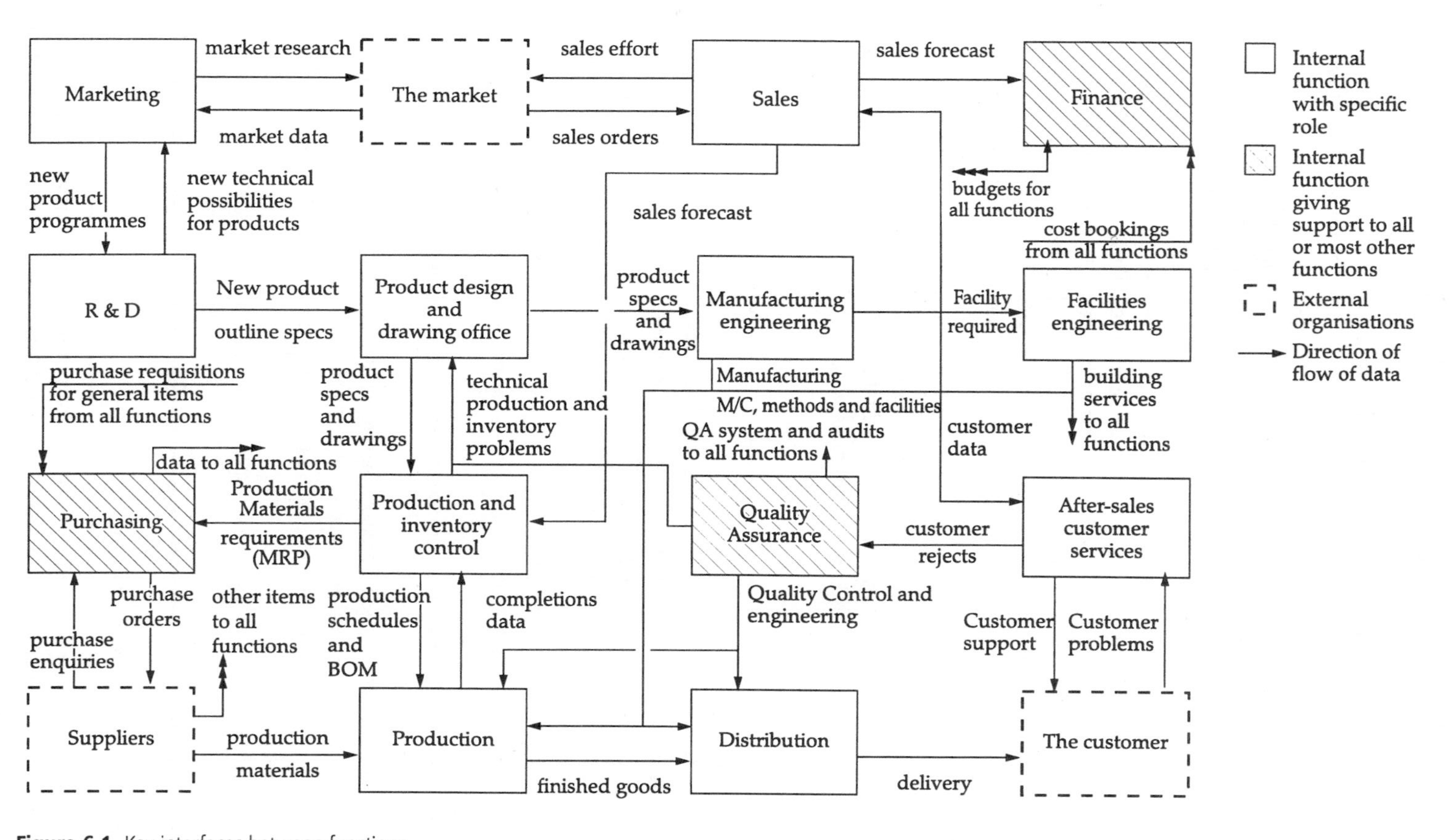

Figure 6.1 Key interfaces between functions

Once a full new product project launch becomes likely, R&D will be given permission and the budget to carry out further work to develop and prove new product principles and features. If this proves successful, approval can be given for full design work and the findings of R&D would be passed on to product design and the drawing office.

The idea of a strictly sequential process of information being passed down the line from R&D to design, design to drawing office, drawing office to manufacturing engineering and quality, is now redundant. Inputs from all engineering departments should be made as early as possible in the research and design processes by the use of concurrent engineering (also known as simultaneous engineering) programmes. These bring people from each of the engineering functions together in a team to consider the performance, quality, manufacturability and costs of the new product through parallel rather than sequential working, see Figure 6.2.

Another major interface for R&D is with external research organisations and academic bodies working in the field of interest. This enables R&D to keep up to date with the latest developments in materials and technology which may have some relevance to the business. Companies will often retain external research organisations to investigate specific issues. The main contact within the company for this work would be the R&D department.

Key interfaces

Key interfaces are with:

- Senior management for the general direction of R&D work.
- Marketing for the outline product parameters to be targeted.
- Design office with the findings of R&D as design inputs.
- Manufacturing Engineering for process capabilities and limitations plus new process developments needed for the new product.
- Quality Engineering for the input of quality considerations and control, particularly on the feasibility of efficient product performance testing and calibration to industry, national and international standards.
- Purchasing Department for the provision of materials and equipment for laboratory experiments.
- External research bodies for additional research on specific problems or to gain information on an area of general research interest for the future.
- Colleges and technical institutes for practical research, desk research and technical information.
- Technical staff of potential suppliers of materials processes and services for incorporation into the product.

Product Design department

Once R&D have provided outline design parameters on new products and approval has been given by senior management to carry on with further

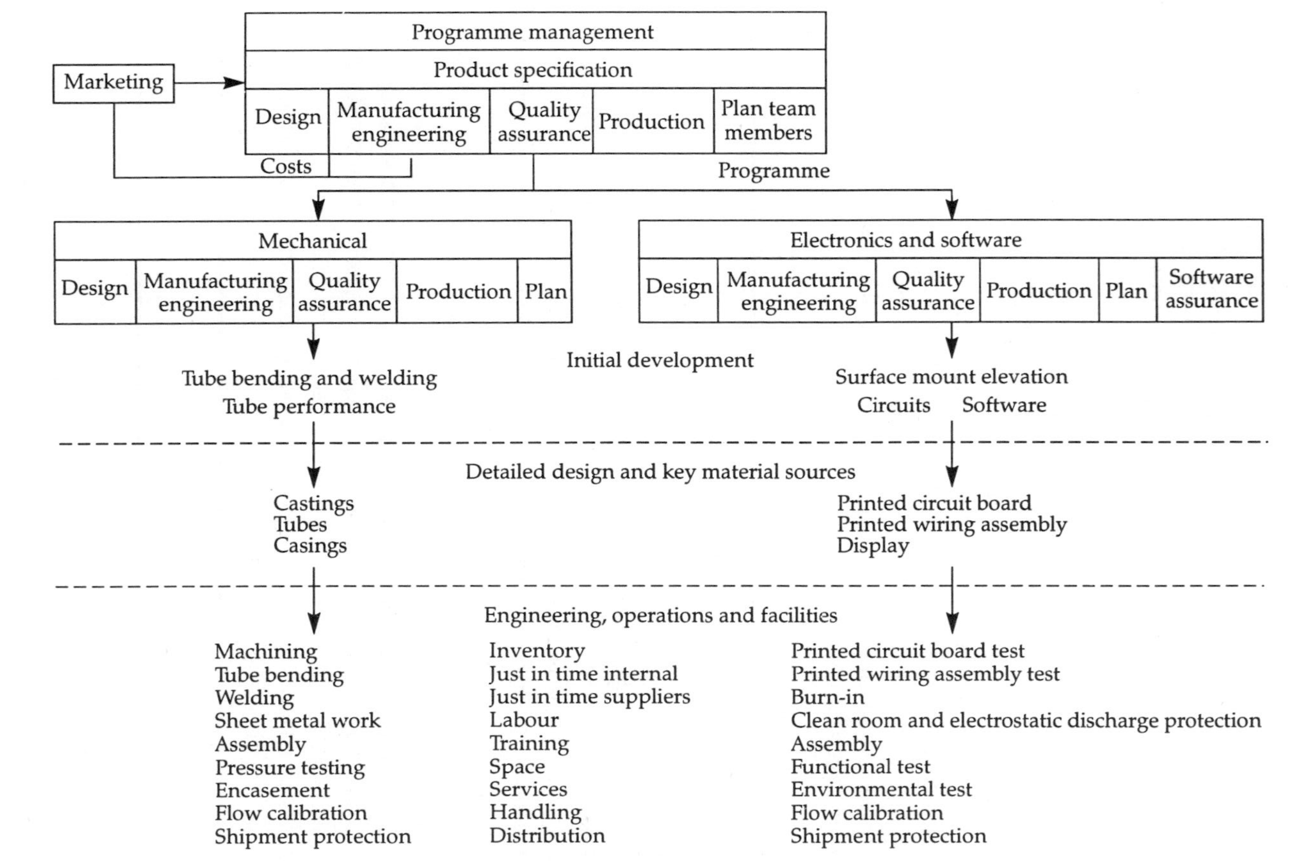

Figure 6.2 Concurrent engineering programme for a combined mechanical and electronic product (Courtesy of Custom Networks Ltd)

work, the design office will undertake full-scale design of the product. On a large programme this in itself may be split into various phases of initial, intermediate and full design, with each needing approval for commencement depending upon the time and cost implications of the design work and the overall importance of the new product to the business. Many design offices will now use computer-aided engineering (CAE) and 3D computer-aided design (CAD) software packages in the design work. Again, this should be done within the orbit of a concurrent engineering programme that takes inputs from manufacturing engineering and quality engineering into the design configuration.

Once the design stage has been completed, the drawing office which may or may not be part of the same department as design within the Product Engineering function, will commence the work of detailed drawing for the sub-assemblies and component parts, parts lists and material specifications. Most new product programmes will be sub-divided by the technologies used and structure of the design. Thus mechanical engineering, electrical engineering and software engineering will all have their own projects and programmes.

Key interfaces

Key interfaces are with:

- Senior management for approval and budgets for design work.
- R&D for the supply of design parameters.
- Drawing office with design layouts for further detailed drawings
- Sales function and potential customers on non-standard product designs and applications.
- After-sales support on problems encountered in product service.
- Manufacturing Engineering for inputs on process capabilities and limitations or the development of new processes in tandem with the new product.
- Manufacturing Engineering (industrial engineering/estimating) for initial costing of the proposed design.
- Quality Control on the acceptability of marginally non-conforming products or materials.
- Quality Engineering for the requirements and possibilities of product performance testing and calibration to industry national and international standards.
- Potential material and component suppliers regarding specifications and supply (in conjunction with the purchasing department).

Drawing Office

On new products

On new products, the drawing office will receive design layouts from the design office. It will then create complete sets of detailed component drawings, sub-assembly drawings and general assembly drawings to manufacture the product. Supporting parts lists, drawing lists, material schedules and material specifications will also be produced.

On existing products

Modifications to existing drawings, lists and schedules are often needed for one of the following reasons:

- Costing saving ideas.
- Change of materials due to loss of availability.
- Change of materials specifications due to change of supplier.
- Modifications to existing designs requested by customers.
- Modifications to existing designs in order to incorporate an improvement perhaps developed for the latest new designs.

Nowadays a substantial amount of design and drawing work will be done on 2D or 3D CAD systems.

Key interfaces

Key interfaces are with:

- Design office for design parameters of new products.
- Customers on their request for modifications to existing designs.
- Manufacturing Engineering with production drawings to enable tooling and processes to be designed and organised.
- Quality Engineering with drawings and material specifications to enable inspection and testing regimes to be set up.
- Production and/or Production Control with drawings to enable manufacturing. Also revisions as they occur and are issued through the drawing office or engineering administration revision control procedures.
- Purchasing for the supply of drawings to suppliers.

Manufacturing Engineering

The term manufacturing engineering is taken to include production engineering, process engineering, industrial engineering, cost estimation, jig and tool design and toolroom. It is therefore the technical arm of production with its inputs into processes, machines, floor layouts, tooling, methods, times and costs. It provides both a technical and an information service to production. As such, it is most often part of the manufacturing function but can also be sited within an integrated engineering function that includes product design and drawing office.

Key interfaces

Key interfaces are with:

- Senior and functional management for approval of general programmes, specific investment projects and budgets.
- Production management for agreement and progressing of manufacturing engineering support programmes to production.

- Production departments in the design, installation, programming and trouble-shooting of production equipment.
- Production departments on the methods, layouts and times of production operations.
- Production departments on the floor layout of equipment and processes.
- R&D and design office as part of concurrent engineering teams working on new products.
- Design office as part of concurrent engineering teams on new products plus trouble-shooting of existing products on behalf of production.
- Design office/drawing office at the interface of CAD and CAM programmes and applications.
- Quality Engineering on the provision of inspection and testing equipment. Also its calibration and usage certification.
- Quality Control on shopfloor problems in the quality of output.
- Facilities Engineering on the supply and routeings of factory services to production departments.
- Facilities Engineering on the installation and maintenance of new equipment.
- Sub-contractors on the supply of tooling to the contractor.
- Equipment suppliers on the investigation and purchase of manufacturing equipment.
- Purchasing on the selection of and contract agreements with equipment suppliers. Plus membership of vendor audit teams.

Quality Assurance (QA)

We will address the interfaces of QA on the basis that the business has certification to the relevant ISO 9000 standard. In this case, QA will liaise with every department in the company to initially set up and then audit their performance to the procedures and controls required by the ISO 9000 standard. Within this environment Quality Control (QC) and Quality Engineering (QE) will pursue their traditional roles. These may also embrace some of the various quality improvement programmes that have been introduced in recent years, one of the most notable being Total Quality Management (TQM).

An enhanced quality culture across the organisation is an important weapon in any modern business. As competition becomes more fierce and industrial and private consumers are not prepared to accept second best, their confidence in the quality of the product is as important as price as long as it is within a certain range. In fact, quality is more important because if the product cannot deliver and maintain the required quality levels over its expected operational life, then it should not be considered for purchase whatever the price.

Key interfaces

Key interfaces are with:

- Quality Assurance
- Senior and functional management on the installation and maintenance of the ISO 9000 standard.

- External certification agencies on their auditing programmes.
- All internal departments on a regular audit of their conformance to the standard.
- Quality Control. This activity embraces the traditional role of the control and inspection of product quality.
- Component production departments on the quality of output via random sampling or full inspection techniques.
- Final assembly and test production departments on the quality and performance of sub-assemblies and the final products.
- Purchasing department and material suppliers upon the quality of incoming materials. This is usually done via a Goods Inwards Inspection Department.
- Stores and inventory control on the protection of goods in storage and shelf-life rules.
- Production, inventory, and design management with periodic (eg. daily, weekly and monthly) quality reports indicating quantities and analysing trends in product quality for each product and department.

  Note. The above inspection operations will operate a non-conformance reporting system that circulates the details of non-conforming products to the relevant departments such as the section producing the product, production and inventory control, manufacturing engineering, design office/drawing office and purchasing department. If discussions among these departments cannot resolve the quality issue, then the product will be scrapped and quality control will issue scrap reports with a similar circulation.
- Design office/drawing office for the consideration of non-conforming product or materials that are borderline to specification and may, in particular circumstances, be accepted for use. Only the design engineers of the product should have the power to grant dispensation because it is only they who have the knowledge of the thinking and design parameters that guided the product specification.
- Manufacturing Engineering on the trouble-shooting of non-conforming product as outlined above.
- Distribution department on the quality of product unit packaging and bulk transportation packaging.
- Distribution on the protection of product in warehouse storage and the shelf-life rules.
- Distribution on the handling and return of reject products from the customer.
- Sales departments on the incidence and handling of reject products from the customer and any associated warranty considerations.

  Note: QC will normally inspect return goods to establish (a) the validity of rejection and (b) the detailed physical/operating condition of the return. An internal report will then be issued to Production and Design office for consideration of the implications of the failure for the design details and production quality.

Facilities Engineering

The term is used here to cover all engineering and technical work undertaken for a business site comprising any combination of factory, offices and warehousing. As such it is also known variously as Works Engineering, Plant Engineering and Maintenance. In this context it includes the engineering of services such as power, gas, water, drainage, compressed air, lighting, heating and ventilation. It may also include building services involved in the maintenance of the site buildings, roads, grounds and perimeter.

The interfaces described are for a large site factory (usually incorporating offices and some warehousing/storage) in which the business owns the site. It is, therefore, responsible legally and financially for all the interfaces listed. In small rented factories, warehouses and offices the landlord would have responsibility for some of the interfaces, particularly those concerning external authorities.

Key interfaces

Key interfaces are with:

- Senior management on planned extensions of the facilities.
- Functional and departmental management on proposed changes to existing facilities.
- Departmental management on maintenance routines for their facilities and production equipment.
- Manufacturing Engineering on the detailed siting, layout and requirements of production equipment.
- Contractors erecting new buildings or installing service facilities such as boiler house, heating and ventilation and other major service supply equipments.
- Services companies such as electrical, gas and the water boards on the main supply and equipment to the site and the metering of the consumptions. In many cases, the cost of services to the site, e.g. the electricity bill will be in the Facilities Engineering's operational budget.
- Local authority environmental officers regarding the outputs and discharges from the site.
- Health and Safety Executive officers on related matters such as quality of air and general ventilation, regulations regarding extraction of process vapours and dusts, machinery guarding and emergency lighting.
- Fire service officers on fire-related safety regulations and practices regarding sprinklers, extinguishers, fire barriers (e.g. siting and fire resistance rating of doors), emergency lighting and emergency exits.
- Fire protection equipment suppliers on sprinkler systems and extinguishers together with their routine maintenance programmes. Note: Many companies employ a Health and Safety Officer who will become involved in all the above considerations affecting the working environment and safety of employees. However, it would normally be the responsibility of Facilities Engineering to physically provide and maintain the necessary equipment and provisions.
- Manufacturing Engineering on the supply of gauges and inspection fixtures and also the inspection and calibration of production jigs, fixtures, tooling and machines.
- Purchasing department on the carrying out of vendor audits upon suppliers.

Production

As mentioned earlier, consideration of factors related to production are included in this section on engineering because daily production activities are more closely involved with the engineering functions that support them than with the commercial functions. This point is debatable, particularly in companies producing a non-engineering product. However, it is still reasonable if marginally correct to include discussions on production in this section.

As has been mentioned previously, the terms production and manufacturing are used interchangeably. In this text the term production refers to the actual making of the product in the production departments and planning of it in a production control department. These are often part of an overall manufacturing function that includes Manufacturing Engineering, Facilities Engineering and Quality Assurance reporting to a manufacturing executive. This arrangement is typical of many businesses but is not universal. It is typical of many small and medium-sized companies producing an engineered product where, in the early days of the business, the initial functional split was between design, manufacture and sales. From this, activities that were not directly concerned with the design or sale of the product, and additionally provided a service to physical production in some way, logically became part of the manufacturing function. For example, Facilities Engineering spends most of its time supporting the factory as its largest user, although the commercial offices will also be supported. Thus Facilities Engineering will most logically come within the manufacturing function.

The same can be said for Quality Assurance (particularly Quality Control and Quality Engineering) although the advent of ISO 9000 has broadened the QA remit. This situation is usually handled by the Quality Assurance manager having a dotted line responsibility to the Chief Executive which provides the manager with the authority to access commercial functions and report serious quality problems anywhere in the organisation. Despite this extension of interest, the majority of quality work is likely to be connected to production activities.

Key interfaces

Key interfaces are with:

- Senior management on the overall manufacturing programmes and budgets.
- Sales for a sales forecast.
- Sales with quotations of delivery times for specific orders and standard lead times for routine production.
- Drawing office on the issue of drawing, parts lists and specifications for production.
- Production Control (a department of the manufacturing function) for the issue of Master Production Schedule, Bills of Material (BOM) and departmental schedules.
- Production Control on the scheduling and progress of jobs. Nowadays much of this will be produced by a suite of computer-based manufacturing planning and control programmes.
- Manufacturing Engineering (Process Engineering) for the design, provision and support of production machines, fixturing, tooling and DNC, CNC and NC programmes.
- Manufacturing Engineering (Industrial Engineering) for the production of methods worksheets, process routeings and standard times. Also process and departmental shopfloor layouts.

- Manufacturing Engineering (Estimating) for production cost estimates and capital expenditure investment appraisals.
- Manufacturing Engineering (Tooling Services) for the supply and repair of fixturing and the supply and sharpening of consumable tooling.
- Quality Control on inspection of product and issues arising from the quality of the output and incoming materials.
- Quality Engineering on the use, calibration and certification of inspection and measuring equipment used in production.
- Quality Assurance on the conformity to control procedures and documentation.
- Purchasing Department on the delivery of production materials.
- Inventory Control or Stock Control (internal departments in manufacturing) on the planning of stock levels, actual stock levels, issues and returns of production material
- Facilities Engineering on the provision and support of services to the factory building and production equipment.
- Distribution department on the scheduling and delivery of finished goods and spares stocks to distribution warehouses.

6.3 Interfaces of commercial functions

Marketing

Marketing is concerned with the issues of which products are sold, how they are sold and how the company will compete in the market place, whilst Sales, which is also concerned with the above in a general way, has responsibility for direct contact with potential and existing customers with a view to gaining and satisfying an order. Most of Marketing's interfaces will be with commercially orientated functions with one major exception which is its inputs into R&D or Product Engineering on issues related to the development of new products.

Key interfaces

Key interfaces are with:

Senior Management for the overall direction of marketing activity.

- Sales function for its projection of sales volumes by market and product.
- Sales function for intelligence feedback from salesmen and what the market (i.e. customers) are likely to want.
- Customers for direct input of market requirements.
- Market research organisations for the conduct of specific research projects.
- Advertising agencies for promotional work.
- Artwork houses and printers in support of the internal production of promotional material.
- Membership of the new product planning committee to agree and prioritise which new products should be worked on.
- R&D or Design office with outline design parameters for the product.
- Finance for product pricing.

Sales

Here it is assumed that the Sales and Marketing functions are separate departments which may or may not report to the same Senior Manager or Director. As we saw earlier in section 4.3, the sales force may be one central unit or split into a number of sales territories or product-centred sales forces. Such variety of internal organisation will create different internal interfaces within the sales organisation but we will concentrate here on its external interfaces with other functions.

Key interfaces

Key interfaces are with:

- Senior Management for the general direction and balance of the sales effort.
- Senior Management, Finance and Manufacturing with sales forecasts and actuals.
- Customers in direct sales activities.
- Customers in indirect promotional work (seminars, conferences, etc.).
- Marketing with intelligence reports from the sales persons and sales managers.
- Marketing with inputs to the consideration of new products and their markets.
- Marketing and Finance on pricing policies – prices for standard products and discounting rules.
- Advertising and promotional staff on the organisation, timing and physical placement of promotional material, e.g. the supply of those materials with distributors and retailers via the area sales person.
- Design office for the design of non-standard products requested by customers.
- Finance and Estimating in support of sales quotations on non-standard products.
- Production or Production Control on delivery dates for customers.
- Distribution on the levels of finished goods to be held in stock and the delivery of products.
- Quality Control on the status, return and inspection of rejected or faulty products.
- After-Sales service on problems arising with the product in service.
- Purchasing department for the supply of operational materials to the function.

Distribution

In some organisations, Distribution is a department within the Sales or Marketing/Sales function. Here it is assumed that it is a separate function charged with the storage, movement and delivery of goods to the customer. As such, it will have warehouses that are separate from the producing factories and a sophisticated transportation network. Whatever the functional arrangements, most of the following interfaces will apply.

Key interfaces

Key interfaces are with:

- Senior Management for the scale and size of distribution operations, e.g. warehousing and transportation networks.

- Sales for the stockholding levels of finished product and spares and agreed delivery network to local distributors, retailers or end user.
- Own factories for the agreed scheduling of factory output to the warehouse.
- Other suppliers for the agreed scheduling of third-party products to the warehouses.
- Air freight and shipping agents for orders for the transportation of product by air or sea.
- UK and foreign customs officials for the granting of import and export licences and payment of the tariffs due.
- Quality Control for the incoming inspection of third-party and own goods plus packaging materials.
- Quality Control for storage condition and for shelf-life rules.
- Quality Control on the handling and return of rejects or faulty goods from customers.
- Purchasing for the supply of packaging and bulk shipment materials.
- Sales, Finance and Quality Control on the return and disposal of redundant stock.
- Manufacturing Engineering for warehouse layout plans and the specification, installation and efficiency of packaging equipment which is often present within such warehouses.
- Facilities Engineering for the building and services support to warehouses.
- The vehicle maintenance facility (if one exists within the company and is not under the control of distribution who would be the main user along with sales fleet cars).

Purchasing

This is normally a central function that supplies a purchasing service to all the other functions within a site. In businesses with multiple sites there will often be a central purchasing function that has small local offices at the various locations. The advantages of this centralisation are the added purchasing power of bulk buying for all locations, better control of the total purchasing expenditure of the business and more conformity in the use of purchasing procedures. In many businesses some form of computer-based materials requirements planning (MRP) or stock control will be a major tool used by Purchasing to trigger orders.

It should be noted that some companies operate a policy that all communications with suppliers must go through the relevant buyer in order to avoid unsatisfactory agreements (either contractural or informal) being made with the supplier by unconnected persons familiar with only their aspect of the relationship with the supplier.

Key interfaces

Key interfaces are with:

- Senior management and Finance for the size and control of the purchasing budget which can be a very high proportion of total costs in a low labour content operation.
- All departments for the order of specialist supplies to each of them, e.g. drawing paper for the DO, machining coolant for the machine shop.
- All existing and potential suppliers for quotations, specifications and delivery schedules to be passed onto the relevant internal departments.

- All existing suppliers with problems arising on delivery, quality or price.
- All suppliers on vendor audits (in conjunction with staff from Quality Control, Design and Manufacturing Engineering) where suppliers' premises are visited to check their technical and productive capabilities and systems of control. This is where having certification to the relevant ISO 9000 standard helps the vendor.
- Credit analysis agencies to ascertain the financial position of potential suppliers.
- Materials management/Inventory Control/Production Control on the scheduling of production materials.
- Quality Control/Goods In Inspection on the rejection of non-conforming material supplies.
- Manufacturing Engineering on the selection and order of production machines and tooling.
- Facilities Engineering on the selection and order of services equipment.

Finance

Finance is dealt with in detail in Chapters 5, 7 and 8. As is stated in Chapter 5, the sphere of financial work that engineers are most likely to encounter is that of management accounting which deals with the internal budgets and costs of the company. Consequently, the majority of interfaces that engineers are likely to see will be in those areas.

Key interfaces

Key interfaces are with:

- Senior management for the agreement of overall financial policy.
- Senior management with the presentation of a master budget for approval.
- All function and departmental managers in the construction of their own budgets.
- Senior and all other managers with periodic reports on their expenditure against budgets and the variances therefrom, for the immediate past period under review and the year to date.
- Senior and all other managers on specific variance reports.
- Sales on matters relating to invoices to customers.
- Customers for the payment of invoices and own credit controls.
- Purchasing Department on matters relating to supplier invoices.
- Suppliers on the payment of their invoices and debit control.
- Production on standard times and production material costs.
- Inventory Control on stock levels and values.
- Manufacturing Engineering on the calculation and entry of standard times and standard costs of production.
- Manufacturing Engineering and departmental managers on capital expenditure approvals and financial calculations.
- Personnel department on employee records for the payment of salaries and wages.
- Outside accounting practices and auditors for the production of company accounts and periodic audits.
- Inland Revenue for PAYE and corporation tax matters and payment
- DHSS for National Insurance matters (NI is paid with PAYE).
- HM Customs and Excise for VAT payments.
- Banks and finance houses for the provision of funds.

6.4 Information Technology (IT)

The impact of IT on businesses has been tremendous, bringing about significant changes in working procedures and facilities in every aspect of commercial and business life. In the very beginning there were just large mainframe computers that were confined to tasks involving large amounts of data work such as national censuses, R&D work and financial programs. The real IT revolution in the business workplace has been facilitated by the introduction of affordable mini- and micro-computers. With the availablity of these relatively cheap machines, a huge software industry has grown up to supply programs that assist just about every business, social and technological activity undertaken in industrialised economies such as ours.

The main thrust of development has been in the United States where the large size of the domestic market has encouraged hardware and software developers to create a constant stream of new products. However, UK and other European companies have also played their part in creating and supplying computer hardware and software at prices that most businesses can afford, even if it is limited to a single desktop computer and printer.

The engineer may use the output from corporate mainframe computers and use its terminals for input of data, but increasingly the engineer is more likely to use mini and desktop computers carrying engineering software and administrative software.

Computer application programmes

Some of the more common computer applications that have been introduced are given in Table 6.1. Figure 6.3 shows the historical relationship of the applications to each other in the construction of a hypothetical full CIM system. Figure 6.4 shows details of the interfaces between computer-aided mechanical engineering software modules. Figure 6.5 outlines the relationships between sales, production and engineering hardwares running off a central computer. Figure 6.6 shows the elements of a shop floor data capture (SFDC) system.

There are many suppliers of the above systems, each claiming a particular edge but the selection of a programme suite or single application needs very careful investigation and an awareness of the system's limitations as well as its advantages. In addition to specialist programmes, general purpose word processing, database, spreadsheet and graphics packages are available to assist administrative work, report writing and the presentation of data.

Many of the applications are run on mini computers using (in the main) variants of the UNIX operating system. However, a large number are now designed to run on desktop computers which we will now discuss.

Table 6.1 Computer applications programs

Abbreviation	Full name
CAD	Computer aided design
CADD	Computer aided drawing and design
CADCAM	Computer aided design/Computer aided manufacture
CAE	Computer aided engineering
CAM	Computer aided manufacture
CAPM	Computer aided production management
CAPP	Computer aided process planning
CASE	Computer aided software engineering
CAT	Computer aided testing
CIM	Computer integrated manufacture
CNC	Computer numerical control
DNC	Direct numerical control
EDM	Electronic data management
MRP	Materials requirements planning
PDM	Product data (or document) management
SFDC	Shop floor data capture

Desktop systems

Hardware has advanced at a rapid pace. The current standard specification offered on desktop computers is typically a 486 or Pentium processor with speeds of 64Mhz or 100Mhz, 32Mb of RAM, 32 bit busing and 500Mb or 1Gb of disk storage. Monitors vary in size and quality of resolution, the larger and better monitors (which are more expensive) being used for graphics work such as CAE and CAD.

These machines are enormously more powerful than machines of just five years ago and it is these increased machine capacities that are allowing ever more powerful software applications to be put into desktop computers and their networks. The two most commonly used designs of machine are the Apple-Mac and the IBM PC and its compatible makes. All now use graphical interfaces (Apple-Mac a long time before the PC and compatibles). All can be connected to a Local Area Networks (LAN) by the addition of a card and networking software.

The most common input devices are still the keyboard and the mouse with light pens and scanners also being used. CD Roms are now superseding 3½" floppies as the preferred form of transportable media. The most common forms of hard copy output are from printers (now commonly laser printers) and drawing plotters. Automatic back-up systems using mini-tape drives are advisable for all but the smallest of users (who can use their floppy disks for the purpose). The installation of uninterruptible power supplies (UPS) is also advisable in order to avoid loss of data and work in the event of mains power failure.

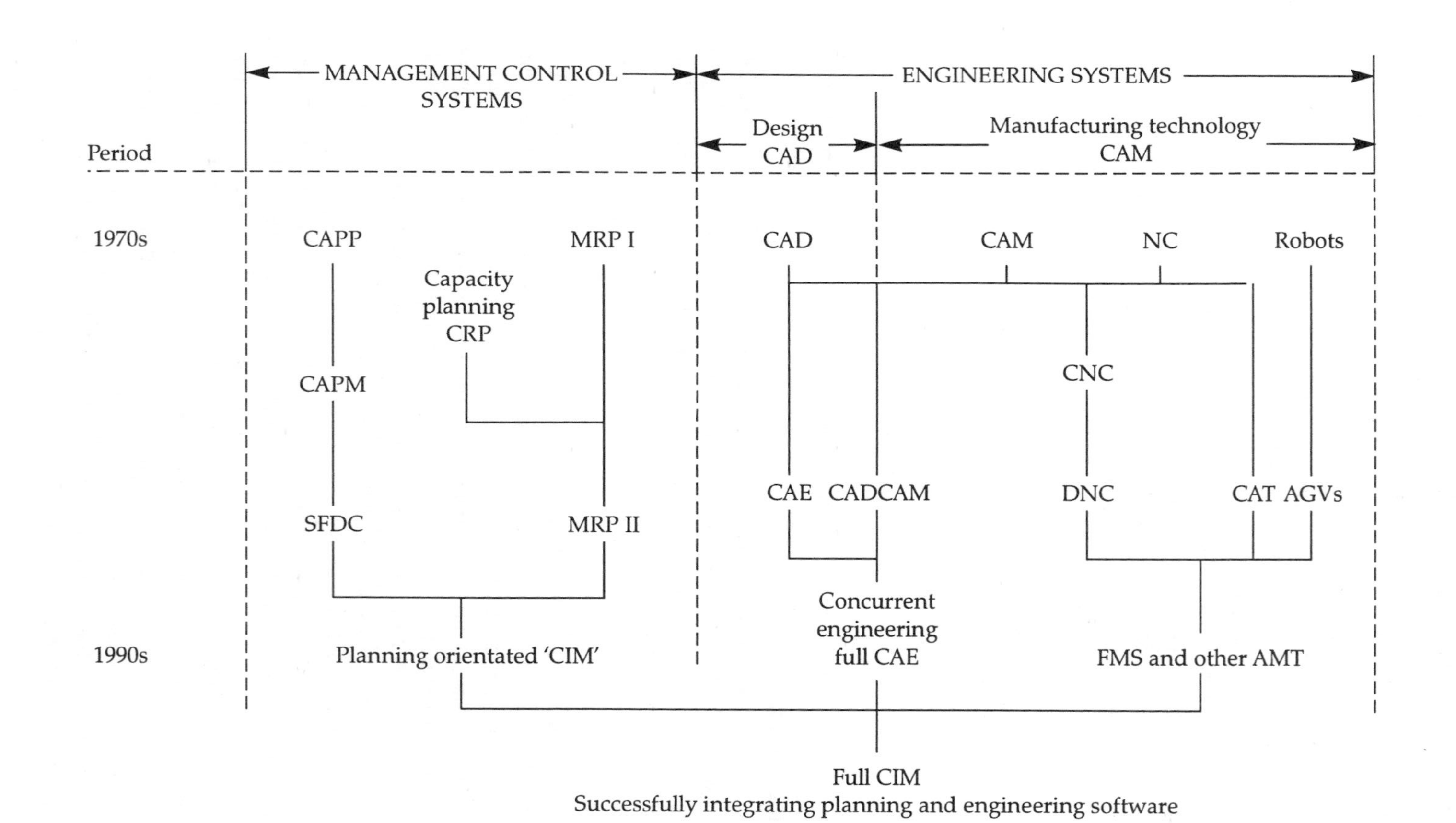

Figure 6.3 Progression of stand-alone computer-aided technologies to full CIM
The figure shows the general application of computer aided technologies by function and over time. It is a simplied statement and does not show the gradual introduction and continuous development of technologies over the time period covered (Courtesy of Butterworth-Heinemann).

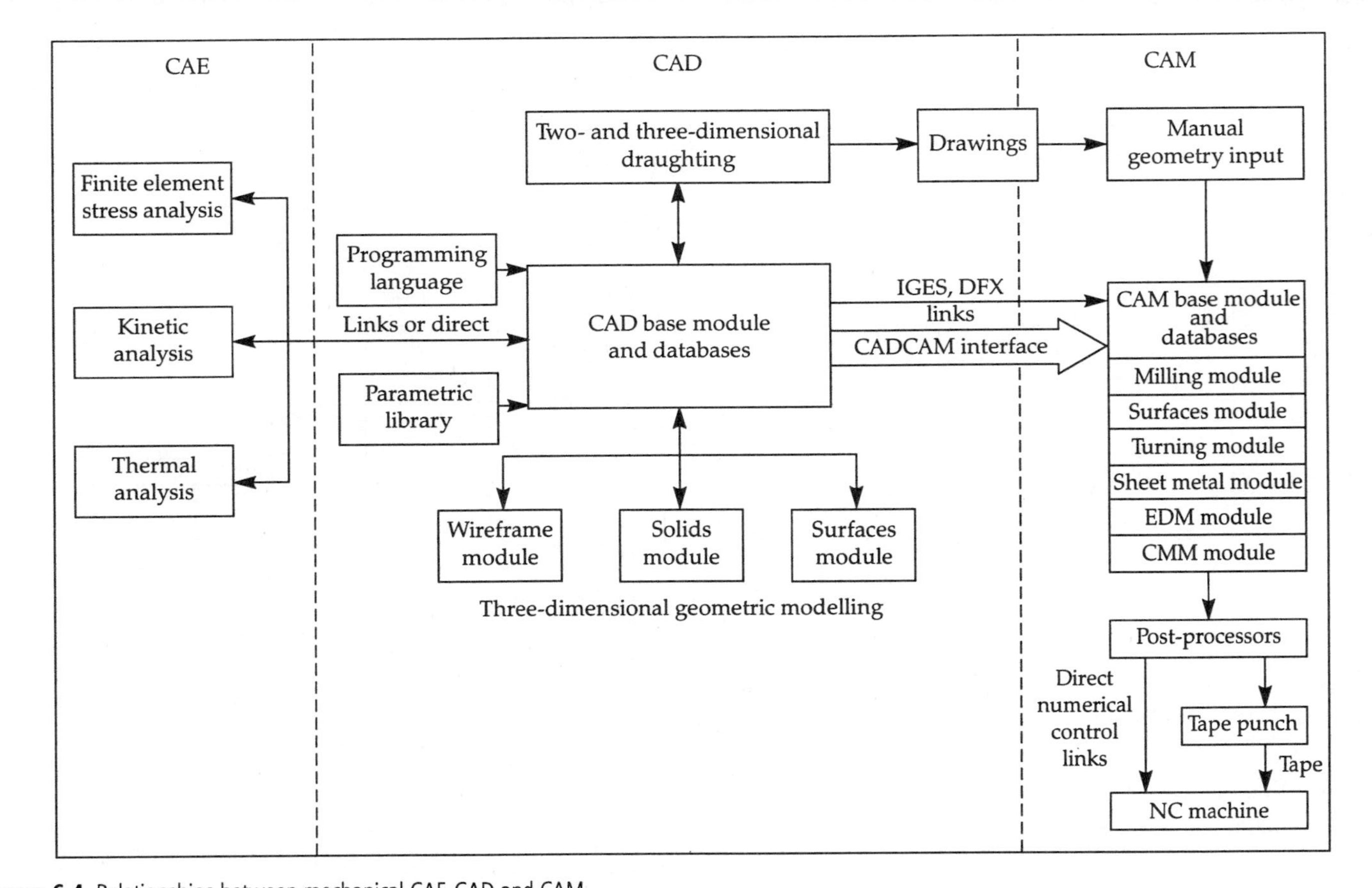

Figure 6.4 Relationships between mechanical CAF, CAD and CAM
EDM = electrical discharge machining; CMM = co-ordinate measuring machines. (Courtesy of Butterworth-Heinemann).

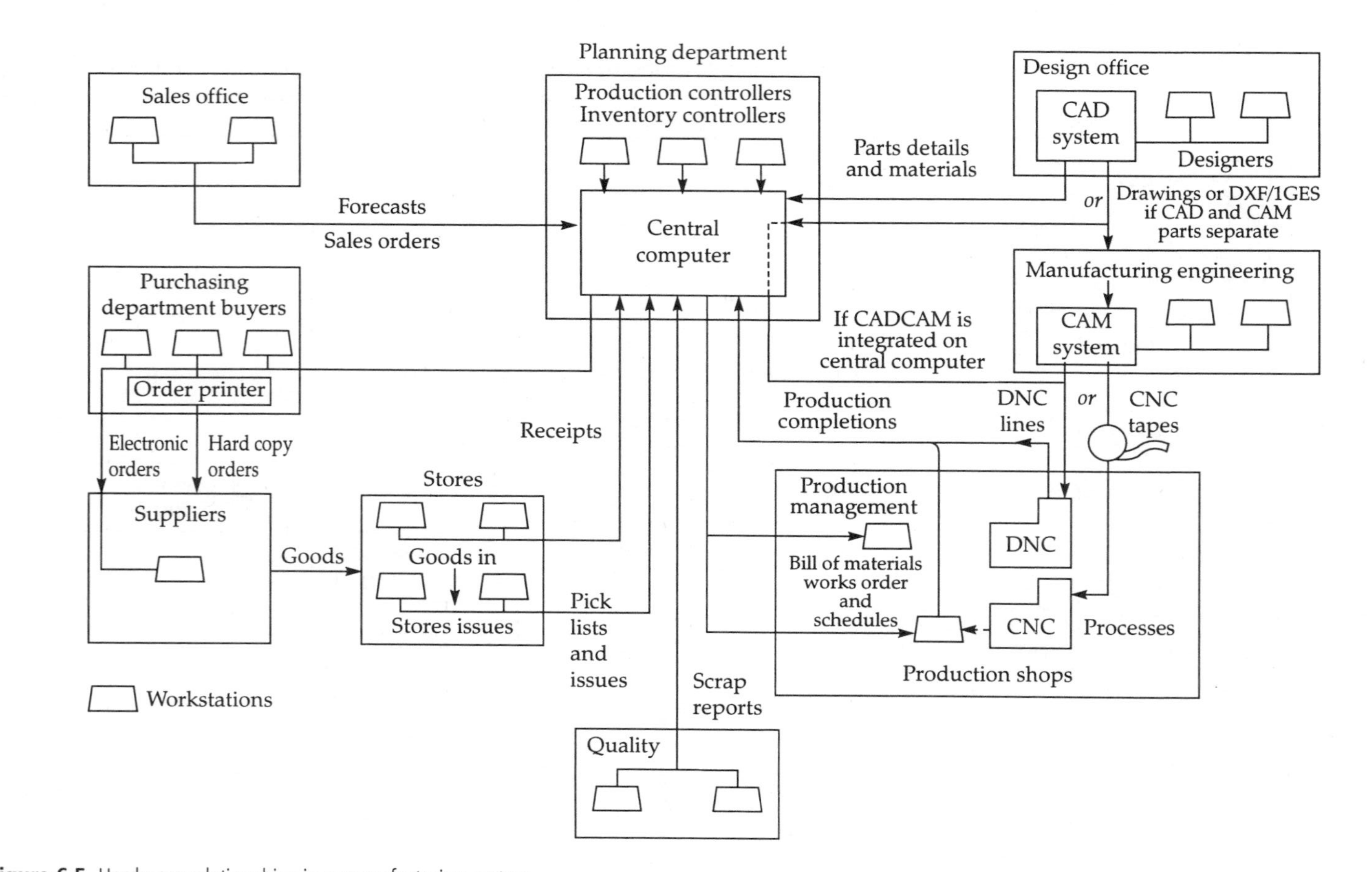

Figure 6.5 Hardware relationships in a manufacturing system
CNC = computerized numerical control; DNC = direct numerical control. (Courtesy of Butterworth-Heinemann).

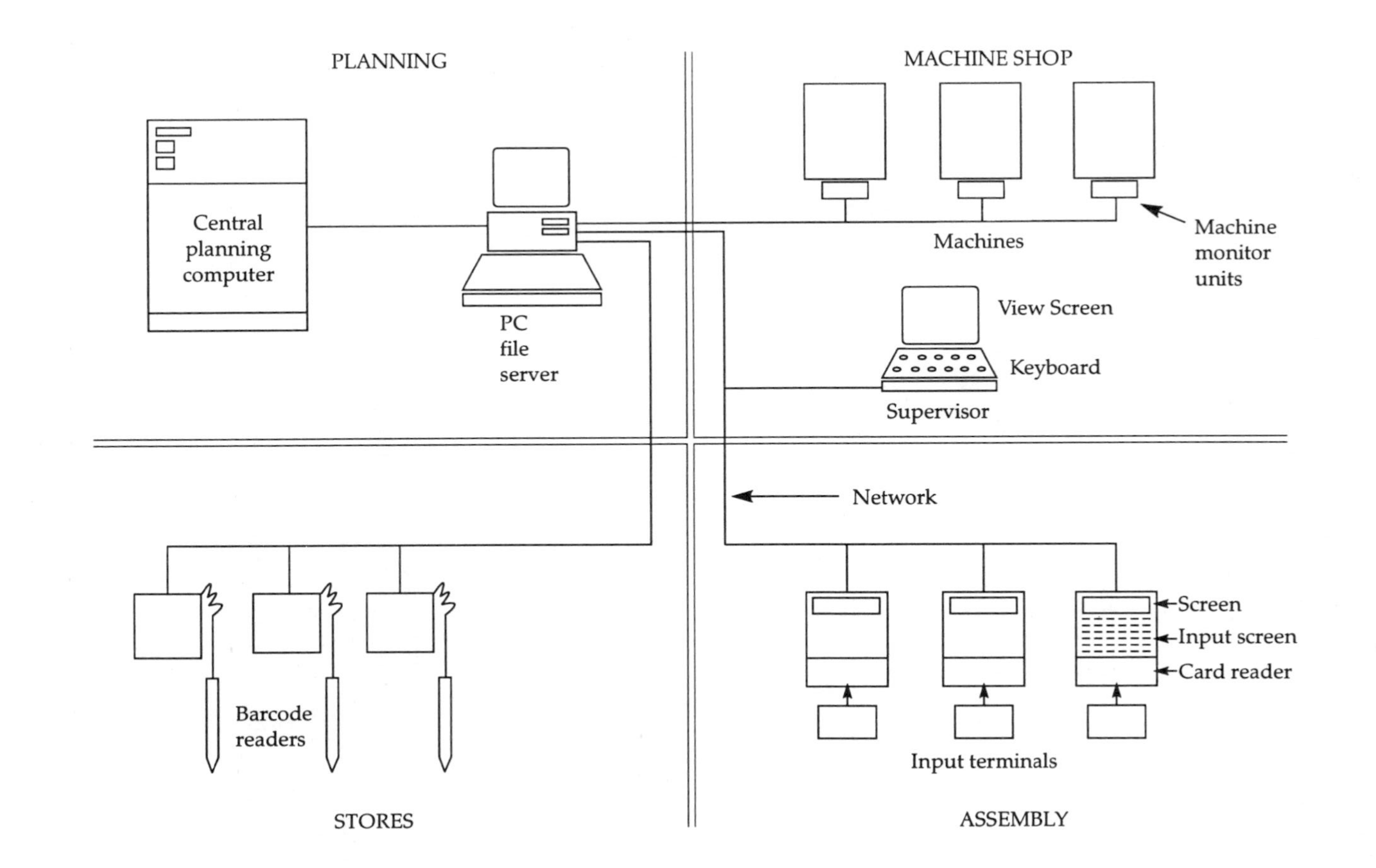

Figure 6.6 Elements of a SFDC system (Courtesy of Butterworth Heinemann)

Modems and communications control software allow inter-site transfers of programmes and data plus the use of electronic mail over standard telephone lines. The advent of the Internet is already expanding access to information held by universities, public bodies, businesses in general and commercial data supply organisations.

Thus, the use of IT is now very common and will increase in every sphere of business activity and communication. It is important therefore for the engineer to be able to understand and use IT to the fullest advantage without coming to regard the medium of IT as more important or interesting than the fundamental tasks it is helping to address. Information Technology is a tool for use, not an end in itself.

6.5 Exercise 6

1. Consider the functions listed below and list the interfaces with four other functions or outside contacts, excluding senior management.
   (a) Research and Development
   (b) Product Design
   (c) Drawing Office
   (d) Manufacturing Engineering
   (e) Facilities Engineering
   (f) Quality Assurance
   (g) Manufacturing
   (h) Marketing
   (i) Sales
   (j) Distribution
   (k) Purchasing
   (l) Finance

2. Name two software applications you would expect to find in the following functions of a modern well-equipped business designing and manufacturing an engineering product.
   (a) Product Engineering
   (b) Production
   (c) Manufacturing Engineering

Part 3

Unit Element 1.3

Applying techniques of monitoring and controlling costs

Overview

A knowledge of costing is essential for a fuller understanding of commercial issues and functions in business. All business activity incurs cost (even the collection and banking of monies received), and in order for a business to make a profit it must have an accurate account of how much is being spent and where and why the expenditure is taking place. This information is needed in order to know:

1. The true cost of the detailed activities.
2. The true cost of each product.
3. The true total costs of the business.
4. Where costs need to be examined because
   (a) the cost is a significant proportion of the total;
   (b) the cost is increasing;
   (c) the cost is decreasing.

The significance of business costs ranges from the total wage bill which can be millions of pounds to the costs of nuts and washers bought for pence.

Costs can be determined by the direct measurement of times and materials taken to complete tasks, recording the charges made by suppliers and apportionments of overhead expenses. For example, the cost of labour can be calculated by measuring time taken for an activity and multiplying by the £ labour cost per hour; the cost of purchased items can be recorded; and the cost of building rents apportioned by calculating the fraction of the rent per square metre of space used by an activity.

Having established the costs of the business, the next task is to manage the costs more efficiently and increase profitability by applying various techniques of cost control. We have, therefore, two major areas of costing activity:

1. The determination of costs under various cost headings.
2. The application of techniques for monitoring and controlling costs.

Chapter 7 will explain the cost headings usually found in a design and manufacturing business and Chapter 8 will explain the techniques of monitoring and controlling costs.

7

Cost headings

7.1 Introduction

There are several heading categories of costs to be examined because costs are defined and used in different ways, each with a particular purpose in mind. The categories are:

1. The element of cost, e.g. wages, materials, rent, maintenance, oil, stationery, insurance and travelling expenses.
2. The type of cost, i.e. direct, indirect, or overhead costs.
3. The behaviour of the costs, i.e. whether they are variable, semi-variable or fixed costs.
4. The techniques of costing used to allocate overhead costs to production. This is known as overhead recovery and includes:
   (a) Total costing (more commonly known as absorption costing.)
   (b) Marginal costing.

7.2 Elements of cost

Every activity undertaken in business incurs a cost of some description. If work is done by staff then a labour cost is incurred in the form of wages paid. If materials are consumed, then material cost is incurred. If people use forms to record their work or electric light by which to do work, then an expense cost is incurred. These three categories of cost element can be used to encompass all the cost that a business normally incurs and we shall now examine each in turn.

Labour cost

The labour costs of a typical batch manufacturing business are usually the largest proportion of total costs, often incurring in the region of 40 per cent to 70 per cent of the total. A large proportion of this is spent in the Production function which often employs the greatest number of people in such a business. It is important, therefore, to know how the labour cost is

being spent and upon which products. In examining the costs of production, the most common basis for calculating the labour cost of individual items is to record the time taken and multiply it by the person's hourly wage rate. In repetitive small component or assembly production this time may be only a few seconds or minutes. Nevertheless, the cost of each item produced will be calculated on that basis.
Example
In the case of a turner producing a small shaft on a lathe:
Time taken for turning shaft = 6 minutes = 0.1 hour
Hourly wage rate of person = £5.00 per hour
Cost of labour for one shaft = time × hourly rate = 0.1 × £5.00 = £0.50

Example

In the case of wiring a large electrical assembly the time taken may be many hours for an engineer who is monthly paid. It is still usual, however, to take a nominal hourly rate × number of hours for the purposes of calculation:

Time taken	= 30 hours
Labour rate	= £1000 per month
Equivalent hourly rate	= £6.66 per hour
Cost of wiring	= £6.66 × 30 = £200.00

Other types of labour cost are incurred by people for whom it is impossible to accurately divide their time. For example, the salary of a machine shop manager cannot be directly allocated against any particular component. These labour costs are treated differently. In this case, the manager's salary will be treated as a different type of cost to be apportioned across all the components produced as a type of cost known as an indirect or overhead cost. We will discuss these in the next section.

Material costs

Material costs are also a significant cost item in a manufacturing company. After all, the essential operation of manufacturing is to convert materials from one form to another through a series of processes. The majority of material costs will be converted in the production processes, and are, therefore, directly attributable to the products that are fashioned from them. In some cases the attribution is easy, for example, a 100mm (4-inch) diameter body casting for an electric motor is only used on 100mm (4-inch) body-sized motors at a rate of one casting per motor. In this case the material cost is the cost of the purchased casting. In other cases attribution is more difficult. For example, the amount of adhesive used to stick components together is less controllable. In this case an apportionment of the cost of the adhesive will have to be made based on the average usage per product or some other parameter.

Expense costs

This category is used to cover all the incidental costs of the business or particular internal activity that cannot be directly allocated to the production of product. Individual elements of expense cost will include charges for:

- rent
- heating
- lighting
- gas
- water
- telephone
- stationery
- insurance
- depreciation
- tooling

Collective areas of expense within the business as a whole would include most office costs, while at the level of production it would include the costs of items or activities used in the general running of the department.

7.3 Types of cost

We have examined the basic cost elements of labour, material and expense costs, but in order to help the accounting treatment of each cost they are additionally defined into three types as follows.

Direct costs

These are costs where all the cost can be directly attributed to the production of a product or service. The following are direct costs:

- Direct labour – the cost of labour directly engaged in production, e.g. machinists and assembly fitters.
- Direct materials – the cost of the materials actually used in the product such as:
  - castings, forgings and bar stock for machining;
  - bought-in components such as fixings and plastic mouldings;
  - complete products like PCBs, transformers, and gear boxes.
- Direct expense – the cost of consumable materials or items such as tooling or printing charges only if it can be said that they will be used solely on the job being costed. This type of cost is more likely to be applied to a particular one-off job or product and not to repetitive production.

The combination of direct labour, direct material and direct expense costs is often referred to as the prime cost of a product.

Example

If a component part is made on machines by drilling and turning operations on a casting, we have three elements of direct cost. In this case there are no direct expense costs.

Direct material – cost of casting	=1.50
Direct labour:	
drilling – 0.1 hour at £6.00/hour	=0.60
turning – 0.2 hour at £7.50/hour	=1.50
total of direct labour	2.10

Thus the total direct cost = £1.50+£2.10=£3.60

Indirect costs (of production)

These are costs that are shared by many products because they arise from general services or materials which are difficult to apportion accurately to an individual production item or activity. Examples of indirect costs include:

- Indirect labour – the wages of production support personnel like supervisors, chargehands, machine setters, store keepers and material handlers.
- Indirect materials – materials used in the production processes but in small quantities from general stocks, e.g. cutting oils, lubricating oils, solder, welding rods, glues, cleaning fluids and materials.

The above are indirect costs of production which are normally added to the direct costs in order to obtain a truer picture of the total cost of producing items. Such costs are normally apportioned to production by calculating a percentage addition to the direct material or direct labour costs. For example, the indirect labour cost of supervision, etc. might add a 20 per cent overhead to the direct labour cost. Storekeeping and material handling might add 10 per cent to the direct material cost. The calculation of the percentages and other methods is discussed in the section on Overhead Absorption.

Continuing our example the ex-production cost would now be as follows:

Example

Direct material	£1.500	
Indirect materials overhead @ 10%	£0.150	
Total material cost		£1.650
Direct labour	£2.100	
Indirect labour overhead @ 20%	£0.420	
Total labour cost		£2.520
Total production cost		£4.170

Overhead costs

These are also indirect costs but are often applied to the cost structure in two distinct stages: factory overhead costs and general business overheads.

The first stage is to add further costs that are incurred by the factory, but which are even more remote from particular items or levels of production. Examples are the heating, lighting, rent, depreciation charges of equipment and repairs for the factory. Another significant factory overhead addition will be total costs of production support departments such as Manufacturing Engineering, Production Control and the works canteen. The organisation structure should be a good guide in this respect. If a support department reports to the manufacturing function executive, then it is reasonable that the departmental cost be included in the factory overhead.

These costs when added to the production cost, give an ex-factory or ex-works cost. In some companies which regard the core functions of Sales and Manufacturing as independent internal trading units, this is taken as the value of output or 'price of its production' that a manufacturing function 'charges' or 'sells' its output to the Sales function.

Example

Total production cost=		£4.170
Factory overhead		
Heating and light @ 1%	£0.417	
Rent of space @ 2%	£0.834	
Overhead departments @ 50%	£2.085	
Total factory overhead		£3.336
Ex factory cost=		£7.506

The second stage of applying overhead costs is to add all the other costs incurred by the business which are costs over and above the ex-factory cost. These are incurred by the functions of the business that do not directly contribute to the production process and normally include the cost of all the sales, marketing, purchasing, financial and administrative functions. The cost of such functions must be added to the ex-factory cost to find the total cost of designing, making, selling and supporting the product to the customer. If the general overheads are not added, then the sale price would be too low and a loss would be made on the product.

Example

Ex-factory cost	£7.506
General overheads	£2.400
Total cost	£9.906

This figure represents the total cost of being in business to produce the drilled and turned casting to the business. It is an arbitrary figure because of the percentage method used to distribute the various levels of the overhead costs. The desired or commercially acceptable profit margin would then be added to obtain the nominal selling price before any discounts or other marketing and sales pricing strategies are applied.

Key points

We should note several things from the example:

- The ex-factory price at £7.506 is over twice the direct cost of £3.60 for materials and labour. The total cost including all business overheads is £9.906 which is nearly three times the direct cost. The increases are an indicator of the significant amount of support and peripheral work that is needed over and above that of actual production to get a product designed and to the market. These scales of increase are common, but in many industries where high ratios of technical or administrative support to production are needed, they can be substantially more.
- The example is a simple arithmetic one used to demonstrate the build-up of costs by using an additional percentage apportionment system for overheads. Details of this and other methods commonly used to calculate indirect and overhead costs and the use of standard costs are explained in later sections.
- The money values are calculated to three decimal places. This is necessary when dealing with percentages and medium volume output to gain the accuracy required in the totals figures. In high volume production four decimal places are often used.

7.4 Behaviour of costs

So far we have examined costs in isolation, as stand-alone items. However, costs are also defined by their behaviour, i.e. what happens to them in relation to the levels of output and the total costs of the business in a financial year.

When money is spent or activities are undertaken which commit the business to an expenditure, it is useful to know if the cost is likely to be repeated or if it is a one-off cost in relation to the production of products. In this respect, costs are classified as being either variable, semi-variable or fixed in relation to the production of output.

Variable costs

Elements of cost which vary directly with changes in the level of output are known as variable costs. This category will include, therefore, most of the direct labour and direct material costs seen earlier in the text because by definition they vary directly with output volumes. Taking the example of the casting, having purchased a quantity for a given price, the cost of one casting will be taken as £1.50 and the cost of 100 castings as £150. The same principle applies to labour costs where the direct cost for the drilling and turning of one casting was £2.10 and the cost of machining 100 castings will be £210.

Semi-variable costs

These are costs which may vary but not necessarily directly with changes in the volume of output because they comprise both variable and fixed elements of cost. Production support activities can fall into this category particularly where work may have to be done at certain intervals of production. For example, manufacturing engineers, quality engineers, maintenance and other support personnel may have to undertake activities after certain levels of production or periods of time have elapsed. In the former case, it could be said that the activity varies with volume but in the latter case it is fixed. A parallel commercial example is in car servicing where servicing is recommended at 12 000 miles or one year whichever comes first; or a telephone bill which is partly a fixed rental cost and partly the variable cost of units consumed.

Fixed costs

Items of cost like building rent remain the same financial commitment to the business regardless of the level of production. Such costs are called fixed costs (and also known as period costs). For example, factory rent might be £10 000 a year regardless of whether zero or a million units of output are produced. The spreading or absorption of this fixed cost over the cost of production will be reflected in the overhead applied to the factory. For example, at a production volume of 1000 per year the unit costs for rent will be £10 000/1000 = £10 per unit. If the output is 100 000 a year, then the cost per unit for rent is £10 000/100,000 = £0.1p.

Figure 7.1 shows the increase in total costs as volume increases and Figure 7.2 shows the reduction in fixed cost per unit.

7.5 Techniques of overhead recovery

This section deals with the apportionment of the overhead costs we have seen above to individual departments and items of production. There are two techniques in common use, namely, overhead absorption (or apportionment) and marginal costing.

Overhead absorption

The terms apportionment and absorption are often mixed in their application but one way of differentiating them is to consider that overheads are apportioned to the costs of whole departments, and absorbed by allocation to individual units of production.

With the overhead absorption method, accountants take the view that all the overheads incurred in providing the facilities for production must be absorbed by the relevant departments in order to know:

- the true total cost of the departments;
- the true total cost of each unit of production.

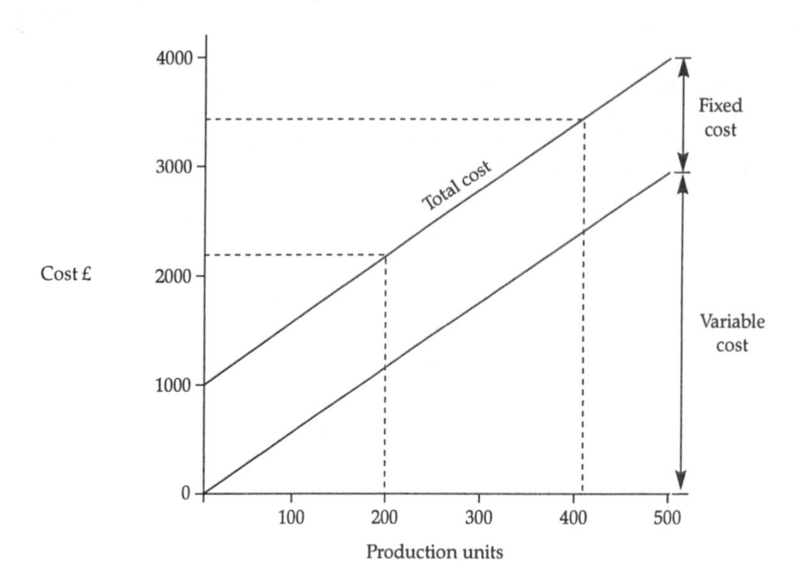

Figure 7.1 Fixed and variable costs of production

Note: If the number of units = 200, the variable cost = £1200, the fixed cost = £1000 and the total cost = £2200. Thus cost per unit is £11.00.
If the number of units = 400, the variable cost = £2400, the fixed cost = £1000 and the total cost = £3400. Thus cost per unit = £8.50.

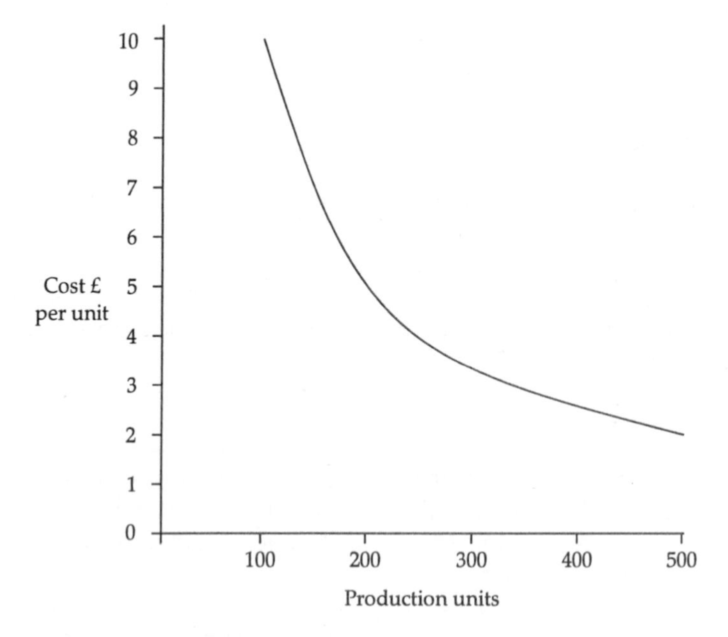

Figure 7.2 Fixed cost/unit production
Note that the fixed cost is £1000. The graph shows how the fixed cost per unit reduces with the increase in the output of units.

We will consider each objective in turn. Overhead costs of the business are normally apportioned in two stages. First, as the overhead incurred by the factory which is added to the prime costs to obtain an ex-factory cost. Second, the other general and administrative overheads, that is the costs of commercial departments are then added to produce a total cost. It may not be thought necessary to distinguish between factory overheads and other overheads in small companies, ones with a simple production organisation producing only one type of product, or where production absorbs the great majority of the overhead. However, in many companies with complex organisations and many products, the separate identification and absorption of factory overhead are thought necessary. Our examples, therefore, illustrate the recovery of factory overhead in such a situation.

Overhead absorption by departments

As we have mentioned above, it is a reasonable and sensible practice to apportion indirect expenses and overheads within the functional structure of management. The manufacturing function would have the costs of Production Control apportioned to it and the sales function would have the costs of the Sales and Administration Office charged to it.

Costs are charged to cost centres which are arranged in a structure that usually mirrors the business organisation structure so that the information can be given to each manager. A typical structure is explained on p.150. All time and material consumptions and purchases are booked against the consuming department cost centre number. This allows an automatic calculation of the totals of departmental expenses in the periodic computer run (normally monthly). These are then amalgamated into the higher level codes so that every level of management has its own financial report.

There are many elements of overhead costs such as indirect labour, indirect material, rent, lighting, heating, stationery, depreciation, insurance, cleaning, telephone charges and the services of such departments as finance and personnel. These are all estimated at different cost levels and rates of expenditure during the year and it would be too crude to just total them all and divide by the level of output to give an overhead per unit of production.

In order to give a more accurate allocation of overheads, the different cost elements are apportioned by using appropriate calculations.

Example: floor area cost basis

In the case of costs such as heating, light, rent, rates and building repairs, these are most sensibly apportioned in relationship to the floor area occupied by the department.

Annual cost of rent, heat, light, etc. = £100 000
Total square footage of the factory = 50 000 square metres
Cost per square metre $= \frac{£100\,000}{50\,000}$ = £2.00 per square metre

Area of machine shop = 18 000 square metres
Annual cost of space to machine shop
$= 18\,000 \times £2.00 = £36\,000$ per annum

Monthly overhead charge on the machine shop budget

$$= \frac{£36.000}{12} = £3000 \text{ per month}$$

Example: number of employees basis

In calculating the overhead apportionment of services to personnel such as the cost of the canteen, drinks dispensing machines, and first aid department, the most appropriate method is by the number of employees in the department.

Total cost to company of such services = £140 000
Number of employees in company = 140
Cost per employee $= \frac{£140\,000}{140} = £1000$ per annum
Number of employees in department = 24
Annual budget for this item is 24 × £1000 = £24 000
Monthly budget charge $= \frac{£24\,000}{12} = £2000$ per month

Alternatively, the apportionments related to personnel in some way may be calculated as a percentage addition to a base wages cost. In the case of the canteen and other employment costs such as the company contribution to personal pensions and National Insurance, the overhead may be applied as a percentage of the department's wage bill.

Example: wage percentage rate basis

Total annual wage bill of factory (£) = 1 000 000
Total annual costs of canteen,
pensions, etc. (£) = 210 000
Additional overhead percentage of
wage bill $= \frac{210\,000}{1\,000\,000} = +21\%$
Department wage bill (£) = 50 000
Overhead absorption rate = +21%
Overhead absorption (£) = 11 500

Overhead absorption by units of output

This method allocates overheads to the units of production, which when summed give the total cost of the product. In principle, if only one unit of production is made, then that unit is said to have absorbed all the overheads that have been spent in the year, i.e. 1/1 or 100 per cent of the overhead expenditure. If 100 units have been made, then they have shared the overheads and the resultant overhead absorption per unit is 1/100 or 1 per cent of the total.

The three ways most commonly used for absorbing overheads by units of output are:

- wage percentage rate;
- direct labour hour rate;
- machine hour rate.

Wage percentage rate

This method is appropriate where there is a significant input of labour to the process and it is applied on a fairly consistent basis rather than a random one, for example, all the workers are operating machines for most of the time. The overhead is then being spread evenly over a consistent situation. If half the workers were not using machines (that consume large amounts of space and electricity), then the overhead would not be spread evenly. Wage rates should be similar across the factory or department. The method of calculating this rate is:

Total factory overhead for period (£) = 10 000
Total direct labour cost for period (£) = 50 000
Direct labour overhead rate $= \dfrac{10\,000 \times 100}{50\,000}$ = 20%

Example

Direct labour cost on batch of 200 = £600
Overhead = £600 × 20% = £120
Total cost of batch = £600 + £120 = £720

Total cost per unit $= \dfrac{£720}{200}$ = £3.600 each

Note: departmental figures for the overhead and labour costs would give a more accurate cost.

Labour hour rate

This method is most appropriate where there is a large element of manual labour (e.g. assembly) applied across the department and the wage rates are the same or very close. The calculation method on a factory-wide basis is:

Total factory overhead for period (£) = 10 000
Total direct labour hours for period = 20 000
Direct labour hour rate $= \dfrac{£10\,000}{20\,000}$ = £0.500

Example

Direct labour hours on batch of 200		= 300
Overhead	= 300 × £0.50	= £ 150
Cost of direct hours	= 300 hours × £4.00/hour	= £1200
Total cost of batch	= £1200 + £150	= £1350
Cost per unit with overhead	$= \frac{£1350}{200}$	= £6.750

Note: a more accurate figure would be obtained if the overhead recovery rate were calculated for each department because both wage levels and overheads consumed vary between departments.

Machine hour rate

This method is appropriate where much of the output is produced automatically on machines without significant labour input or alternatively for groups of machines doing like work on a repetitive basis. In these cases the time that the machine is utilised is a good basis for the allocation of overhead. The overhead to be absorbed is usually that for each department rather than the whole factory because of the great variation of machines that need to be considered – a general factory overhead applied proportionally as the basis for a departmental machine rate would be very inaccurate. The method of calculating this rate for each department is:

Total departmental overhead for period (£)	= 2000
Total departmental machine hours for period	= 3000

Machine hour rate $= \frac{2000}{3000}$ = £0.660

This figure should be added to the cost of production for every hour that the machine runs.

Marginal costing

This is an alternative accounting technique to total overhead cost absorption by the units of output. The technique takes the view that having established a factory and its attendant fixed overheads like rent and heating, then the production of an additional unit of output will only incur additional direct and other variable costs and not any further fixed costs. Therefore, in considering the profitability of producing more output, the fixed overhead cost should be excluded because it undervalues the additional contribution to profit made by the additional unit of output.

Example

Contribution = Sales revenue – (direct costs + variable overheads)

Sales revenue of additional 1000 units			= £10 000
Direct costs labour and materials		= £4000	
Variable overheads			
Tooling	£1500		
Other expenses	£500		
Total		£2000	
Total marginal costs			£6000
Contribution to profit = 6000 – 2000			= £4000

Thus, with the fixed overheads already creating a fixed financial commitment regardless of the level of output, the production of a further 1 000 units will contribute an additional £4 000 to the company profit.

Marginal costing is used in make or buy decision-taking. In this case the marginal cost of in-house production is compared with the purchase price. Again, the principle taken is that whether the product is made in-house or not, the fixed costs have already been incurred and are therefore irrelevant to the decision. Variable overheads for items such as additional tooling and production semi-variable costs would need to be included in the variable overhead element of the marginal cost.

The technique is also useful in the high volume production of goods by automation, where the cost of running the equipment and factory facilities are constant and only a small amount of direct labour is involved in unit production. In these cases the greatest additional cost incurred in producing more units will be their direct material cost, which in mass production is normally small in comparison to other costs. However, if the other costs are effectively fixed, i.e. already 'paid for', this marginal addition of material cost may be the only significant detractor from the additional profit generated

Another advantage is that marginal costing produces a constant product cost upon which sales can base their prices regardless of changes in the volume of output in the factory. The constant cost figure is maintained because the contribution is not recalculated to adjust for fixed overhead absorption every time the level of output changes.

Conversely, fixed overheads cannot be excluded from the pricing structure altogether, otherwise they will not be recovered from the customer, and a loss will be made. However, the fixed costs can be included as a final general overhead to obtain an original base price.

Marginal costing is only applicable for marginal changes in output levels that involve no changes in the fixed assets or costs. If an increase of, say, 10 per cent in output necessitated the purchase of costly capital equipment or extensions to the factory, then marginal costing would not be appropriate because of the significant increase in fixed overheads and their recovery rate.

7.6 Exercise 7

1. Give three reasons for a business maintaining accurate accounts of its financial transactions.

2. Name the four categories of cost and costing.

3. What are the three major elements of cost that are found in the manufacture of a product? Give an explanation of how they relate to each other.

4. Which cost elements would you expect to be included in the indirect overheads of a factory?

5. Explain the difference between direct and indirect costs. How are indirect costs included in the cost calculation for a manufacturing operation?

6. A production operation on a component A has the following costs:

Direct material	= £1.20
Indirect material overhead	= 15% of direct material
Direct labour hours	= 0.25
Direct labour cost	= £5.25 per hour
Indirect labour overhead	= 20% on direct labour

Calculate:
 (a) the total cost of the operation;
 (b) the percentage of the cost attributable to labour costs;
 (c) the percentage of the cost attributable to material costs.

7. A second production operation on the same component A incurs the following labour:
 Direct labour time = 0.4 hour
 Direct labour rate = £6.00 per hour

Calculate:
 (a) the cost of this operation;
 (b) the total cost of the component after the two operations.

8. The overheads for the factory in which component A is produced are:
 Rent and rates = 1% of prime cost
 Service department's charge = 45% of prime cost

Calculate the prime cost, the rent and rates and the service overhead. Also calculate the ex-factory cost.

9. If the general overhead for administration and sales is 60 per cent and the target profit margin is 25 per cent of total cost, calculate the selling price of component A.

10. Give two examples each of variable and fixed costs.

11. Draw a nominal graph showing how such costs build up against unit volumes.

12. Describe the basic principle of overhead absorption.

13. How are overhead costs allocated on a floor area basis for department B? Calculate the monthly budget allocation for overhead from:

    Total area of 30 000 square feet = Total cost £86 000 per annum
    Area of department B 4000 Square feet

14. Company X has 250 employees. The annual cost of support services to the employees is £35 000 p.a. Department A has 30 employees, department B has 25 and department C 15 employees. Calculate the monthly budget allocation for each department on a cost per employee basis.

15. If in the next year company X increases its staff to 300 employees and the costs of employee support rise by 15 per cent, what will be the new monthly figures for departments A, B and C who each take on 2 new employees?

16. Business Y has an annual wage bill of £520 000. The overhead costs the employees is £100 000 p.a. Calculate:

    (a) the wage percentage rate of overhead allocation;
    (b) the monthly overhead absorption for department A with an annual wage bill of £30 000;
    (c) the monthly overhead absorption for department B with an annual wage bill of £27000.

17. A moulding shop runs six machines on a 24-hour, 28-day per month basis. The overheads incurred in a 28-day month are:
    Rent and rates = £1000
    Power and heating = £ 300
    Machine minding = £3000
    Calculate:
    (a) the machine hour rate;
    (b) If one machine produced 30 000 mouldings in a month, what is the effective overhead per moulding?

18. Explain the principle behind the use of marginal costing and its advantage over the overhead apportionment method when calculating the cost and profitability of additional production.

19. Product X has a sales revenue of £25 000 over a quantity of 1000 units in one month. Variable direct costs amount to £8000 and variable overheads 22 per cent of direct cost. Calculate:

    (a) the marginal cost of a batch of 500;
    (b) the contribution to profit.

8

Techniques of Monitoring and Controlling Costs

Having examined the various types of cost, we will now look at the techniques used to monitor and control them. There are two types of technique – budgetary control and operational control systems.

8.1 Budgetary control

Budgetary control is a technique in which all the activities of the company have a financial budget set against them at the beginning of the year and as the year progresses the actual costs incurred are recorded and compared with the budget. If the actual results coincide with the budget, then the management can be reasonably sure that up to that moment the business will achieve its financial and operational targets. Any differences (called variances) between the actual results and the budget can be seen and should be investigated to establish the reasons. Corrective action can then be taken to bring the operation back in line. Budgetary control is therefore a powerful monitoring tool for use by management in the drive to meet the annual financial and operational targets.

The technique is applied at all levels in a business, by setting a master budget for the whole company which is then broken down successively to functional, departmental and section budgets. Thus, the total expenditures of the business are related in a logical and detailed manner with the sub-budgets usually aligning to the organisation structure (i.e. the management responsibilities) of the company. Figure 8.1 shows a budget structure.

Types of budget

The financial control of a company has to be considered from several points of view and types of cost. As a result, a number of budgets are produced from the basic data mentioned above. These will include:

- master budget;
- administration budget;
- cash budget;
- departmental operating budget;
- capital expenditure budget.

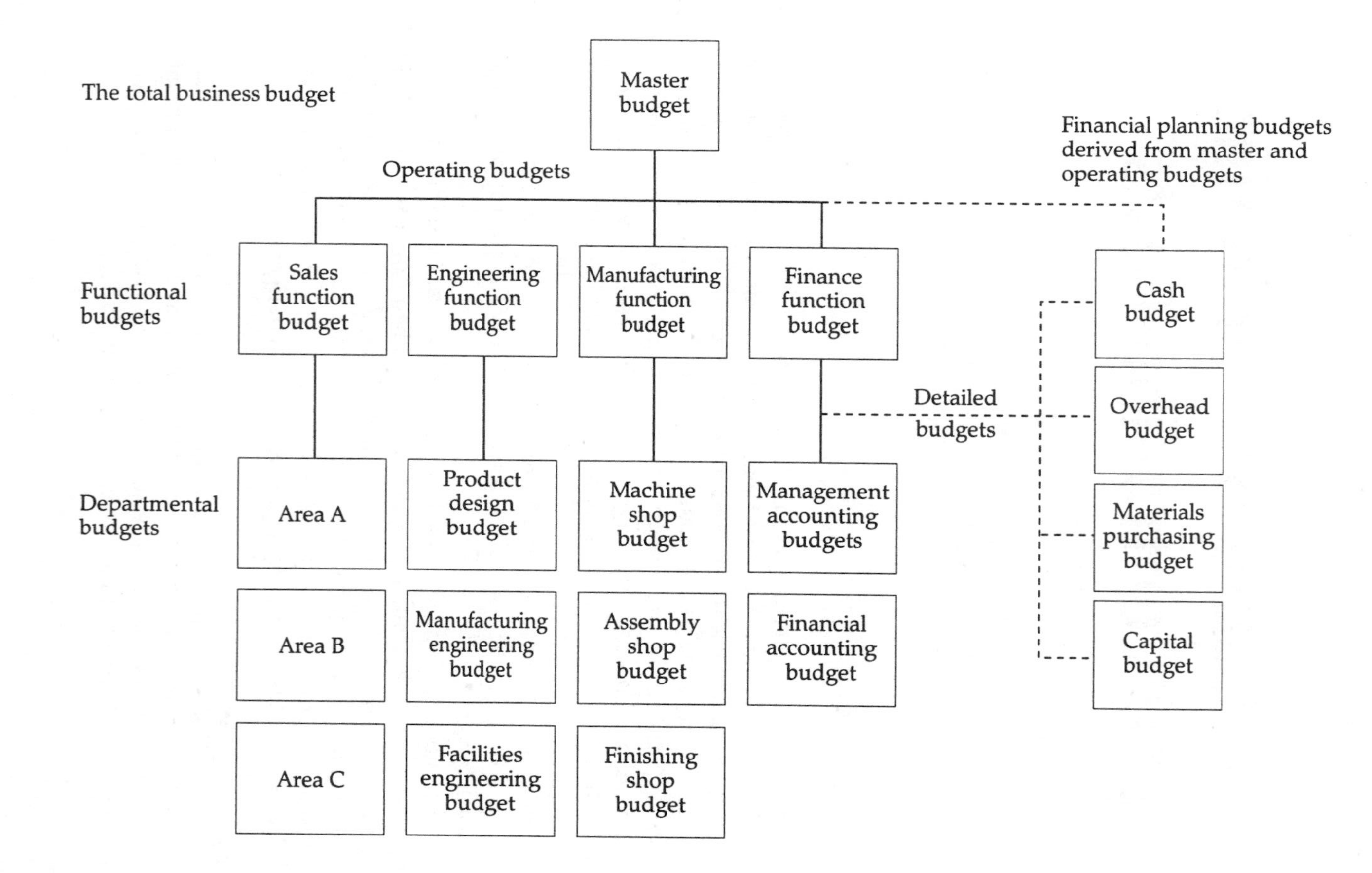

Figure 8.1 Budget structure

The first three listed are normally the concern of the accounting function and the engineer is more likely to meet budgeting issues in relation to the departmental operating and capital expenditure budgets.

Departmental operating budget

The departmental budget will include all the expenditures of the department that occur on a regular basis. These are judged to be part of the overall cost of supplying the product to the customer (regardless of whether they are direct, indirect or overhead expenses), and are treated by the accounting system as costs to go into the profit and loss calculations of the business. Labour, materials, incidental expenses and overhead expenses would fall into this category. In order to instill a common system of control for all departments and functions each type of cost is given a unique cost code number. In many accounting systems it is a four-digit numeric code, for example, direct labour may be given the code of 1010 and stationery 6110. This makes it easy for the accounting software programmes to summarise the totals of a particular type of expense (e.g. direct labour or stationery) for the whole business and also to identify the departments spending the most on each item. The functions, departments and sections of the business become cost centres and are also given a number system so that costs can be accurately booked by staff and then added up by the computer programme. Table 8.1 shows an example of a departmental operating budget for a machine shop which is planned in four three-month quarters. Table 8.4 gives as example of a cost centre number code system.

Key points

- The annual operating budget with its monthly reports is the major financial tool that managers have for the control of their departmental finances throughout the financial year.
- All costs incurred by the departments are recorded via systems of cost code numbers for each cost element being booked to cost centre numbers for each budgeted function, department or section.
- These systems keep the booking and calculating operations completely numerically based and avoid errors occurring through different descriptions being used for the same cost element or department.
- The monthly budget report of each cost centre shows the budgeted values and values actually spent for every cost code for the month under review, plus the year-to-date performance and the variance of the actual figures from the budget. The variances alert managers to the areas that need attention because the actual performance is straying from the budgeted (i.e. planned) targets.

Capital expenditure budget

Capital expenditure is money spent on the purchase of fixed assets such as machinery, service plant and buildings. As such, it is not regarded as part of the operational profit and loss account but is an item for inclusion in the balance sheet. This is because capital expenditures provide an asset with a

Table 8.1 Annual operating budget shown in quarters (Jan–Dec 1997)

Units of output: item	cost code	Qtr 1(£)	Qtr 2(£)	Qtr 3(£)	Qtr 4(£)	Total
		5000	5000	5200	5500	20700
Basic labour	1010	50000	50000	52000	57000	209000
Overtime	1020	5000	5000	4500	3000	17500
Incentives	1030	5000	5000	4500	3500	18000
Total direct labour	1000	60000	60000	61000	63500	244500
Indirect labour o/head	1100	5000	5000	5200	5700	20900
Total labour costs	1200	65000	65000	66200	69200	265400
Direct materials	1310	25000	25000	26000	27500	103500
Scrap	1320	500	500	520	550	2070
Indirect mats o/head	1350	2500	2500	2600	27500	35100
Total material costs	1300	28000	28000	29120	55550	140670
Expenses						
Consumable tooling	3000	1250	1250	1300	1500	5300
Stationery	5010	100	100	100	100	400
Travel	5020	300	300	300	320	1220
Maintenance	5030	1000	1000	1040	1100	4140
Depreciation	6000	2000	2000	2000	2000	8000
Floor space	7010	1500	1500	1500	1500	6000
Lighting and heating	7020	700	700	700	700	2800
Total expenses cost	9000	5600	5600	5640	5720	22560
Grand total		98600	98600	100960	130470	428630

continuing value which adds to the worth of the company, whereas departmental expense payments are ultimately consumed by suppliers and employees and leave no residual asset value in the company. The return in the latter case is the cash payment for the product by the customer.

It is important, therefore, from the accounting point of view, to have a separate budgeting and recording system for items of capital expenditure. Manufacturing engineers and facilities engineers are most likely to be involved in the preparation and control of capital budgets. In some companies the Manufacturing Engineering Manager or equivalent has responsibility for the capital budget because the majority of capital expenditure is on plant and equipment planned, purchased and introduced by Manufacturing Engineering. Table 8.2 shows a capital budget typical of a factory with between 400 and 1000 employees engaged in a mixture of fully automatic, semi-automatic and manual production operations in machine shops, finishing shops, assembly and test operations. Budget reporting may be similar to the departmental expense system with reports on budgeted, actuals and variances for the month under review and the year-to-date performance. Justification of capital expenditures was dealt with on p. 87.

Table 8.2 Capital expenditure budget (Jan–Dec 1997)

Item	Project type	Qtr 1(£)	Qtr 2(£)	Qtr 3(£)	Qtr 4(£)	Total
Machine Shop						
CNC Lathe	Cost saving	100000				100000
CNC Machining centre	Cost saving		160000			160000
CNC Grinding m/c	Expansion				80000	80000
Auto – 1" bar	Expansion		65000			65000
Cleaning bath	Replacement			2500		2500
Hand drill grinder	Replacement			1200		1200
Racking for tool storage	Necessity	500				500
Total		100500	225000	3700	80000	409200
Assembly Shop						
Assembly cell (P243)	New product	20000	5600	10000		35600
Brazing machine (P243)	New product		10000			10000
Circuit board tester (P243)	Cost saving	8500				8500
H123 Assembly m/c	Replacement			23000		23000
PCB soak oven	Expansion				8000	8000
Screens for welding area	Health and Safety	2500				2500
Total		31000	15600	33000	8000	87600
Manufacturing Engineering						
Toolroom tool grinder	Replacement			27000		27000
Toolroom spark eroder	Replacement		15000			15000
2 – CAM workstations	Expansion	2500		2500		5000
Total		2500	15000	29500		47000
Drawing office						
OA size printing m/c	Replacement				4000	4000
1 – CAD workstation	New Product		2500			2500
Total			2500		4000	6500
Quality Department						
Co-ordinate measuring m/c	Cost saving			95000		95000
Replacement gauges	Replacement	500	500	500	500	2000
Reference test oven	Necessity		6000			6000
Total		500	6500	95500	500	103000
Stores						
Pallet racking – castings	Health and Safety	5000				5000
Shelving – small items	Expansion			1200		1200
Fork-lift truck	Replacement		8000			800
Total		5000	8000	1200		14200
Grand total		139500	272600	162900	92500	667500

Stages of budgetary control

Budgetary control has four stages:

1. Forecasting.
2. Preparation and approval.
3. Monitoring.
4. Adjusting.

Forecasting

Budget forecasts are prepared with data coming from two directions. First from the top where senior management will indicate the policy and operational guidelines within which the budget must sit. Many companies have long-range forecasts of their sales and operations which are updated every year – with next year's budget then becoming the first of the next five years. Overall business plans and budget guidelines from senior management will be based on the long-term forecast plus the results of the current year. In many companies the sales forecast will be the primary document determining the forecast levels of activity of all the company. In addition, other influencing factors will also determine elements of the budget. For example, if there is to be a freeze on employing more people and it is anticipated that the company could only afford a 3 per cent pay rise in the coming year, a general guideline will be issued to managers stating that 'labour costs must be limited to last year's level plus 3 per cent'.

Second, managers and supervisors will prepare their budget forecasts based on historical data and experience. Thus if a supervisor has been able to introduce a cost-cutting measure in the latter part of the current year, then next year's budget forecast should show a lower cost for that item.

Preparation and approval

Budget preparation is an iterative (repetitive) activity which can take many months in large and complex organisations. For example, it is not unusual for a budget that runs from January to December to be in preparation from August of the previous year. Budgets should NOT be set entirely by accountants but must include the operational experience of the managers and supervisors who will be responsible for achieving the budget targets set.

The annual budget can start at any time of the year but will usually be at the start of a calendar month. The year is divided into control periods which are normally one calendar month (12 per year), a four-week month (13 per year) or four 13-week quarters. Weeks are also used for planning and budgetary purposes with the weeks numbered from 1 to 53. This is a very useful tool in all planning because it avoids the inevitable confusions of using actual dates.

It is necessary to divide the year into control periods because neither incomes or expenditures are smooth throughout the year, and managers and the accountant controlling the cash flows of the company need to plan their activities to reflect the variances during the year.

A typical programme for the preparation of an annual budget might be

1. Senior Management issue budget guidelines based on the five-year plan, the sales forecast for the next year and their current policy decisions on other influencing factors, such as government fiscal policy e.g. interest rates.
2. Sectional and departmental supervisors prepare their budgets from the guidelines.
3. Functional managers consolidate the section and departmental budgets into one budget for their function and submit these to the budgeting section of Finance Department.
4. Finance department amalgamate all the function budgets into a business master budget and submit it to senior management along with all the supporting functional budgets.
5. The senior management review these and make general amendments to make the master budget conform to the five-year forecast and the other influencing factors mentioned above. In many cases this will mean reducing the budget amounts forecast by managers and supervisors as there is a natural tendency for them to be optimistic in their planning. In other cases, it may involve increasing budget allocations because of policy intentions known to the senior management but not at that time known to functional or departmental managers.
6. Budgets are returned to all departments with instructions to make the detailed amendments necessary to achieve the general amendments.
7. Budgets are re-submitted. Budgets are amended and re-submitted once more to the approvals process.
8. Senior management approve final budget for the coming year.

In large companies the budget forecasting and preparation programme can take several months. The author knows of companies that start the budget programme in August in order to be able to work to an approved budget starting on 1 January of the following year.

The approved budgets become a major control item in the operation of the business throughout the year and are used to continuously monitor the progress of the departments against their plans.

Monitoring

The monitoring of budgets is achieved in the following way.

1. Each department is issued with a copy of its own budget which both the manager responsible and his or her subordinate and superior managers have agreed during the preparation and approval process.
2. The cost of all expenditures is collected through a cost recording system that collects data on costs incurred and time spent in the departmental activities.
3. At the end of each control period (e.g. every month), a budget report is issued showing the actual cost, budgeted cost and variance between them. This is normally done for both the control period under review

(i.e. last month) and also the totals for the year to date. For example, in July a budget report will be issued showing the figures for June and the totals for January to June inclusive. Table 8.3 shows such a report. In addition to these departmental budget reports, the accounting function may issue specific variance reports. These summarise the variances of specific types of cost over functions and/or the whole company. The types of report often produced are:
(a) Material cost variances.
(b) Labour cost variances.
(c) Overhead cost variances.

Adjustments

The budget reports give operating managers and the senior management a snapshot of the company's financial and operational performance to date and therefore an opportunity to take corrective action where necessary. If the action required is relatively minor, then the original budget figures will be kept and the variances against them shown. If the variances lead to significant changes in the activity levels or directions of a department, function or worst of all, the company, then a new budget with revised target figures may have to be prepared and approved.

Cost centres

Many computer-based accounting systems are available and most use a numbering system to identify cost centres which correspond to management reponsibilities in the organisation structure (*see* Table 8.4). This provides managers with a statement of the costs incurred by their departments that is usually presented in the monthly budget report. Each functional manager will have a top-level report which is a summation of the lower level reports of his/her subordinate managers. This system then progresses down the management structure with succeeding levels to the lowest budget reporting level.

8.2 Inventory control

The nature of inventory

Inventory is another name for stocks of materials, parts and finished goods. Consequently, inventory control is a technique for the regulation of the levels and value of these items. The cost of inventory can be a high proportion of the total product cost, typically ranging from between 40 per cent to 80 per cent of the ex-factory cost, depending upon the nature of the product.

The high proportion of materials within the finished product means that, at any one time, a considerable amount of material will be held in various physical states within the business. Consequently the value of materials held at any one time is a significant figure and the primary objective of

Table 8.3 Monthly budget report for June

Cost item	Cost code	Month budget	Month actual	Month variance	Month %	YTD budget	YTD actual	YTD variance	YTD %
Basic labour	1010	50000	48597	1403	2.8	300000	306967	–6967	–2.3
Overtime	1020	5000	3258	1742	34.8	25000	24584	416	1.7
Incentives	1030	4800	4325	475	9.9	20000	18735	1265	6.3
Total direct labour	1000	59800	56180	3620	6.1	345000	350286	–5286	–1.5
Indirect labour o/head	1100	5000	4800	200	4.0	30000	30645	–645	–2.2
Total labour costs	1200	64800	60980	3820	5.9	375000	380931	–5931	–1.6
Direct materials	1310	25000	24456	544	2.2	145000	156389	–11389	–7.9
Scrap	1320	500	432	68	13.6	2900	3981	–1081	–37.3
Indirect mats o/head	1350	2500	2445	55	2.2	14500	15639	–1139	–7.9
Total material costs	1300	28000	27333	667	2.4	162400	176009	–13609	–8.4
Expenses									
Consumable tooling	3000	1200	1094	106	8.8	7500	7863	–363	–4.8
Stationery	5010	100	110	–10	–10.0	600	490	110	18.3
Travel	5020	300	150	150	50.0	1250	1387	–137	–11.0
Maintenance	5030	1000	856	144	14.4	7500	7093	407	5.4
Depreciation	6000	2000	2000	0	0.0	12000	12000	0	0.0
Floor space	7010	1500	1500	0	0.0	9000	9000	0	0.0
Lighting and heating	7020	700	650	50	7.1	4200	3978	222	5.3
Total expenses cost	9000	5600	5266	334	6.0	34550	33948	602	1.7
Grand total		98400	93579	4821	4.9	571950	590888	–18938	–3.3

YTD = year to date

Table 8.4 Example of a four-digit hierarchical numbering system for cost centres

	Level 1	Level 2	Level 3
Manufacturing Function	1000		
Machining Dept		1100	
Turning Section			1110
Milling Section			1120
Assembly Dept		1200	
Mechanical Assembly			1210
Electrical Assembly			1220
Production Control		1900	
Scheduling			1910
Progressing			1990
Engineering Function	2000		
Product Engineering		2100	
R&D			2110
Drawing office			2120
Manufacturing Engineering		2200	
Process Engineering			2210
Industrial Engineering			2220
Others			2290

inventory control is to reduce this value without disrupting production or service to the customers through unavailability of stock.

Types of inventory

Inventory is usually classified in the following manner:

- raw materials;
- purchased components and sub-assemblies;
- components produced by the factory;
- sub-assemblies produced by the factory;
- finished goods.

The flow of these inventories through a business is usually a continuous process in which all materials are also classified as being either:

- stock;
- work in progress (WIP);
- finished goods.

Stock

Stock includes all materials that are held in stores waiting to be used in the product. It would include raw materials such as lengths of steel bar, purchased components such as castings or plastic mouldings, purchased assemblies such as switches or electric motors and sub-assemblies produced by the factory. As stock, these items are held in a store ready for issue to various stages of production and are usually under the control of a stores or inventory manager.

Work in Progress (WIP)

This is usually taken as the materials that have been issued to the shop floor and are being processed or waiting between processes. WIP is normally under the control of the production department to which it was issued.

Finished goods

This is completed and tested product awaiting delivery to the customer or possibly in regional warehouses prior to delivery to the customer. In a small company, finished goods might remain under the control of the Production or Factory manager. In larger companies with established distribution organisations, finished goods are usually under the control of the Distribution or Sales function.

Inventory costs

The cost of inventory is made up of three elements:

- the purchase cost of the material;
- the administrative cost of placing a purchase order;
- the internal cost of storing and handling the material throughout its progress from Goods-in to despatch.

Figure 8.2 shows how these costs are incurred over the material cycle.

A problem arises in costing materials in that any one type or part number may be bought at different times, quantities and therefore prices. For example, one type of casting in a stores' stock may comprise of quantities bought at £1.00, £1.10, and £1.20 each. So when trying to value the total stock held or apply a direct material cost to the materials issued and used in production (see Chapter 7, Elements of costs), problems arise as to which cost to apply.

Three methods of determining the cost to be applied are in common use.

- First in, first out (FIFO). Here it is assumed that the material first delivered to the stores will be the material first drawn from the stores and therefore the cost of the first material can be used.
- Last in first out (LIFO). Here it is assumed that the last material delivered to the stores is the first to be used, so in this case the cost of the last material purchased is used.
- Standard material cost. This is a cost that has been calculated so that it can be applied to all valuations and direct material costs. The method is very commonly used and provides a standard material cost from which the cost of one product to another, one process to another and one batch to another can be compared. When added to the direct labour costs concerned it gives a standard cost of production.

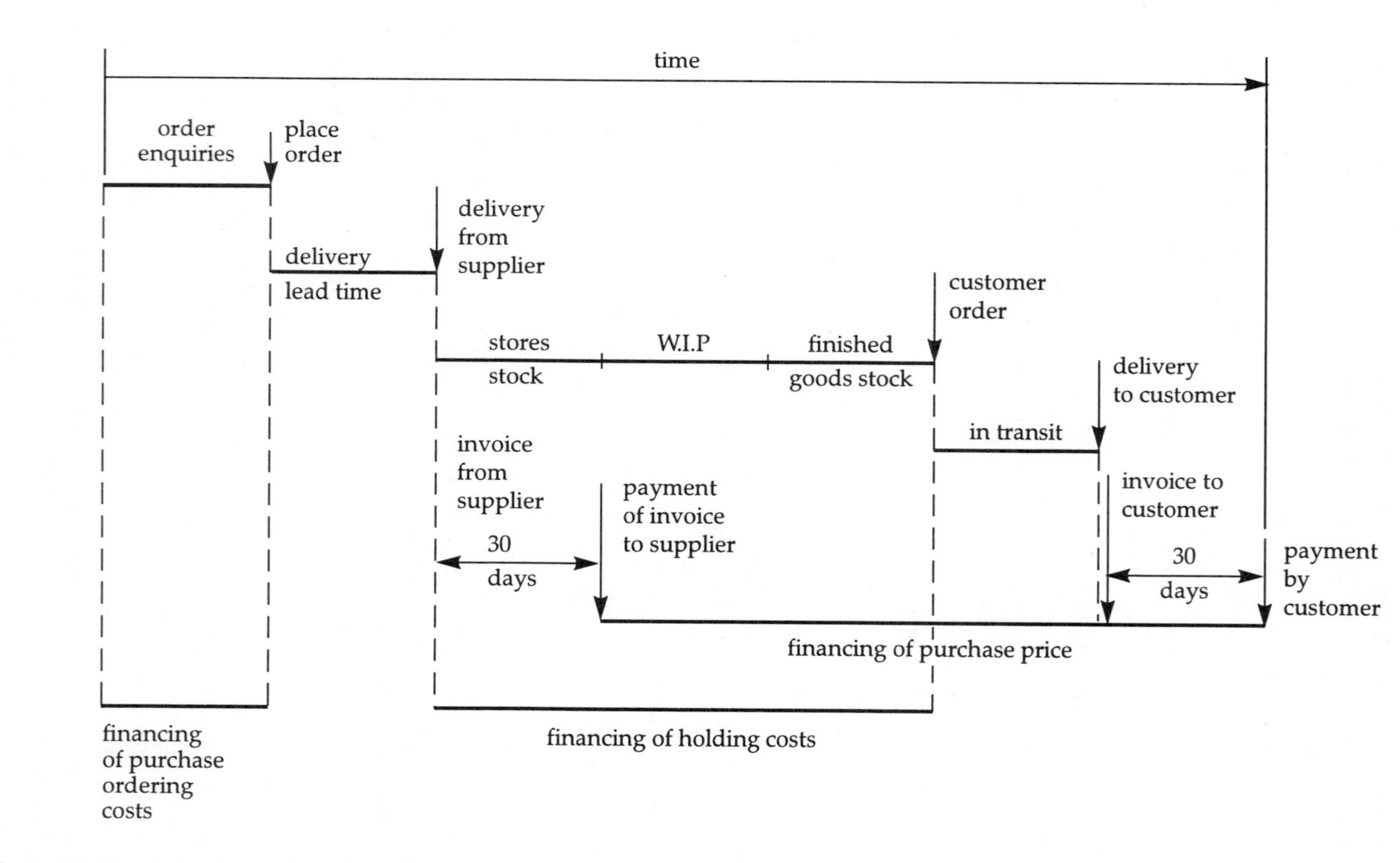

Figure 8.2 The physical flow and financing of inventory

None of these methods gives an accurate cost to any particular single piece of material used in production. Such accuracy can only be achieved if the cost is tracked individually for, say, a casting so that its inventory cost can be allocated to the finished product or unit that it goes into. This may be done for one-off orders where the parts are unique but in repetitive batch and mass production, costs calculated by one of the above methods are considered to be sufficiently accurate for accounting records and management decision purposes.

Inventory levels

The main method of keeping inventory cost as low as possible is to keep the inventory levels to a minimum that gives a balance between the level of stock held and the levels of demand. Figure 8.3 shows the stock holding and demand for an item and what happens to both over a period of time. The chart shows that the level of stock will range from a peak at the time of a delivery to a minimum just before the next delivery – assuming there has been a typical pattern of demand in the meantime.

If delivery could be arranged to be at the instant after point A (zero) and before point B (the next demand), such an inventory situation ranging from a peak to zero would theoretically be acceptable. However, given the problems that can arise in any production situation, where delivery and demand can fluctuate in quantity and time of happening, the planning of inventory levels on the basis that they will reach zero at frequent intervals carries great risks. It is normally prudent to arrange deliveries so that a minimum level or buffer stock is maintained, below which the stockholding will not fall under normal

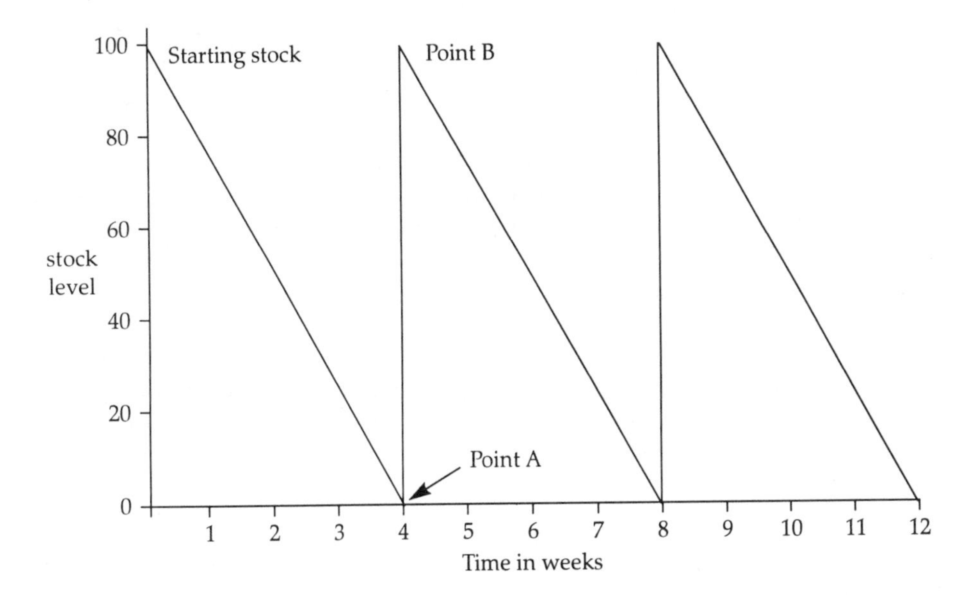

Figure 8.3 Inventory levels
Note that demand is 25 per week or 100 per 4-week period.

demand patterns. Figure 8.4 illustrates the inventory cycle if a buffer stock of 20 units is allocated to a demand pattern of 25 units per week.

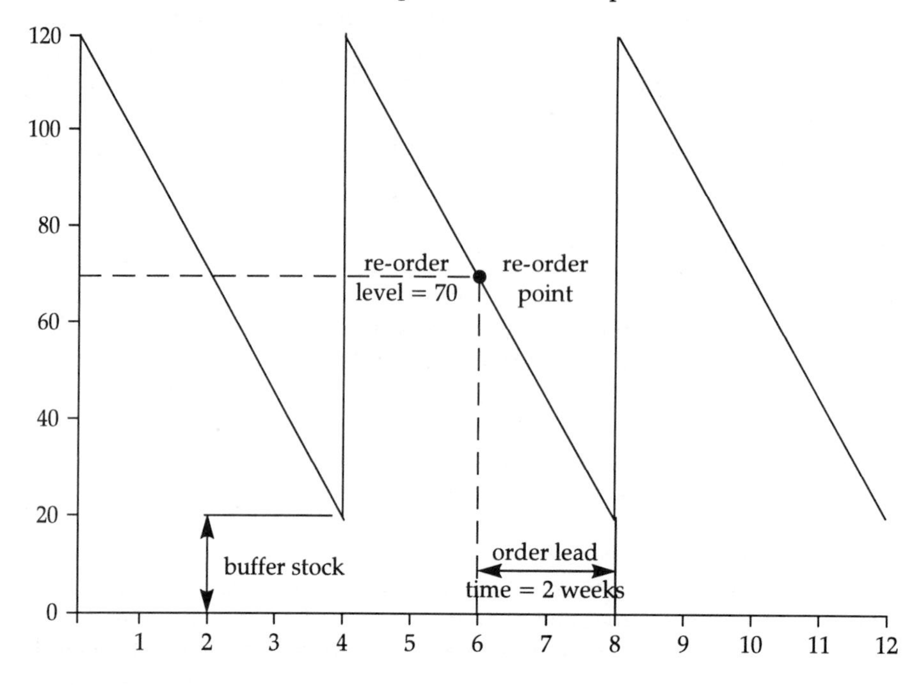

Figure 8.4 Inventory buffer stock and re-order point: demand 25 per week

There is, however, another consideration to add to inventory planning and that is the time delay encountered between ordering an item and receiving it. This is known as the lead time, hence it is essential that new stock is ordered at a point that allows for the delivery lead time. Figure 8.4 also shows the re-ordering point set to allow for a two-week lead time within a four-week demand period.

If a business is to control its inventory costs effectively, the stock levels and ordering point required to meet the forecast demand must be calculated for all materials but the very cheapest items that can be economically purchased in bulk, e.g. standard washers or screws costing a few pence each. The effort of making inventory calculations on a regular basis throughout the year for most of the materials consumed is considerable. In order to handle this work, most modern businesses have computer-based inventory control programs which do the calculations automatically every week or month. The programs are often integrated in a suite that also contains purchasing, production planning and financial control packages.

Just in Time (JIT)

The inventory control factors described above are essentially concerned with the control of batches of material. However, the materials are often processed one at a time or in small numbers. This regime means that in a batch of 100 castings, 99 will be waiting either to be processed or as processed parts. If the batch size could be reduced, there would be less stock sitting around at an inventory value of £x waiting to be processed and £x + £y process cost after being processed. From this point of view the ideal batch size is one.

The Japanese technique of Just in Time production is based on this principle, and has been widely introduced in Western economies in an effort to reduce inventory costs by reducing batch values and reduce total lead times for production by avoiding the batch waiting times discussed above.

The major aims are to reduce material levels and costs and production lead times and this is done by adopting the principle that the ideal batch size in any operation is one. Materials will then only be called forth and consumed at the time that it is needed to provide one unit of output, hence the term Just in Time. The introduction of JIT was not merely a matter of adopting another control technique but a method of inducing a re-arrangement of production processes and equipment and attitudes in order to allow single or very small batches of items to be produced. The difference between JIT and the traditional methods may be summarised as the traditional method makes parts in batches against forecast demand and puts them into stock. This creates high inventory values and long production lead times. JIT makes an item or very small batches only as they are wanted by actual orders. This keeps inventory levels low and shortens production lead times.

JIT working requires more flexible working practices and reduced set-up times if the potential inventory cost savings are not to be nullified by the cost of the increased number of set-ups. JIT working also demands other improvements if the flow of production and service is to be maintained. These are:

- improved quality – there is no buffer stock or production time available to replace defective materials;
- improved and guaranteed deliver – there is no buffer stock to cover late deliveries.

These demands apply equally to internal suppliers, i.e. the prior process as well as to external suppliers.

With both JIT and traditional production, businesses are improving quality and delivery from external suppliers by increasing liaison and joint planning of the supply chain. Formerly, businesses just ordered items and left the supplier to handle the order within its own business with no accountability to the customer of what went on. Increasingly, customers and suppliers favour more integrated planning and access in both businesses in order to establish one continuous supply chain running through the two organisations. The advantage of this for both businesses is the reduction of confusion, misunderstanding and delays, plus avoiding having to take costly and time-consuming corrective action.

8.3 Forecasting

Forecasting is the process of predicting the future over some time period by projections and extrapolations of the historical and present situations. It is undertaken by the business as a whole and by all functions and departments within it. Forecasting is necessary in order to provide a starting-point and objective for all detailed planning of activities within the business. It is, therefore, a key tool in planning and controlling the expenditure and cost of the company.

Periods of forecast

Forecasting is usually applied over three time frames:

- long-term forecasting;
- medium-term forecasting;
- short-term forecasting;

Long-term forecasting is normally applied for periods of two years or more with five- and ten-year forecasts used to determine:

- Strategic marketing and sales forecasts.
- Strategic manufacturing forecasts.
- New product developments.
- New factory, warehouse and office sitings.
- Long-term contracts with customers and suppliers.
- Financial forecasting including investments, takeovers and amalgamations.

Medium-term forecasting is normally applied in annual or bi-annual time frames to determine:

- planning of sales and production targets over the next two years.
- The introduction and market launch of new products in that period.
- Investigation and approval of proposed capital investments for introduction in the period.
- Financial planning of budgets and cash flows for the financial year to come.
- Agreement of prices on upcoming contracts with customers and suppliers.

Short-term forecasting applies to monthly, quarterly and annual time frames to determine:

- Detailed sales and manufacturing output plans.
- Consequential schedules for labour, materials and overhead consumption.
- Departmental budget forecasting, actual performance and variance reporting.
- Ordering of materials against schedules produced by inventory control/MRP.

- Detailed planning of delivery schedules to customers.
- Detailed planning of engineering support programmes and projects.

Forecasting techniques

The projection of intelligence and opinions

This is particularly applicable to marketing and sales forecasting where what the customer wants and is likely to do has to be anticipated. Inputs of information and opinion from sales people, retailers, wholesalers will help to form a marketing or sales forecast of what is likely to happen and what the response should be. In addition, results from market research exercises on customers or the general public can be processed statistically but in the final analysis are mostly based on opinions taken rather than documentary evidence of what has actually occurred in the past.

Thus, forecasts produced from intelligence and opinion are highly subjective and carry a great risk of inaccuracy if there is no additional input of data into the forecast.

Historical data analysis: simple trends

This technique will take historical data and project it into the future using the same mathematical trend and relationship as seen in the past. Figure 8.5 shows a graph using this approach. The graph shows what the forecast based on simple trend projection would be for a product with an output of 32 in January, 36 in February, 26 in March, 34 in April, 42 in May and 38 in June.

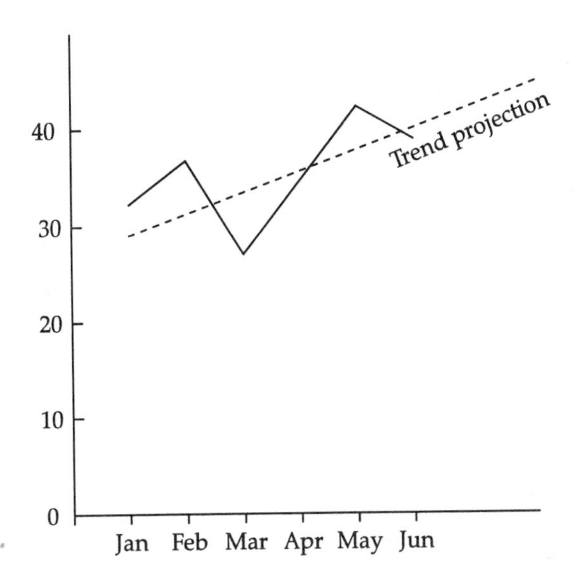

Figure 8.5 Forecast based on simple trend projection

Moving averages

This uses the technique of establishing an average for the historical data over a reference period and applying that in a forecast. For example the average sales over the last three months (months 1 to 3) might be used to determine the forecast sales in month 4. Similarly in month 5, the average sales in months 2 to 4 will be taken and in month 6 the average sales of months 3 to 5 will be taken. Figure 8.6 shows a graph of moving averages.

Weighted moving averages

One of the short-comings of simple and average data-based forecasts is that they do not take account of any variations in the importance and criticality of particular items of data. For example, in a moving average of actual sales, the latter months are more important than the earlier months. It would be prudent therefore to assign a ranking if importance or weighting to the monthly figures.

This is done by applying a weighting factor to each element of data. The weighting factor is expressed as a fraction or decimal of one (1) so that the total of all weighting factors will equal unity.

Example

Data element	A	B	C	Total
Weighting factor	0.2	0.3	0.5	1.0

Example

The figures for the past six months' sales are 100, 120, 120, 110, 130, 140. The simple average is 120 per month. However, unless there is a known seasonal or operational fluctuation to be taken into account, the latter months are more significant. Therefore, a weighting factor should be applied. On a three-month moving average, the weighted figures would be:

	Jan	Feb	March	April	May	June
Sales Units	100	120	120	110	130	140
Jan.–Mar.						
Weighting factor	0.2	0.3	0.5			
Weighted units	20	36	60			
Weighted average		116				
Feb.–Apr.						
Weighting factor		0.2	0.3	0.5		
Weighted units		24	36	55		
Weighted average			115			

	Jan	Feb	March	April	May	June
Sales Units	100	120	120	110	130	140
Mar.–May						
Weighting factor			0.2	0.3	0.5	
Weighted units			24	33	65	
Weighted average				122		
Apr.–June						
Weighting factor				0.2	0.3	0.5
Weighted units				22	39	70
Weighted average					131	

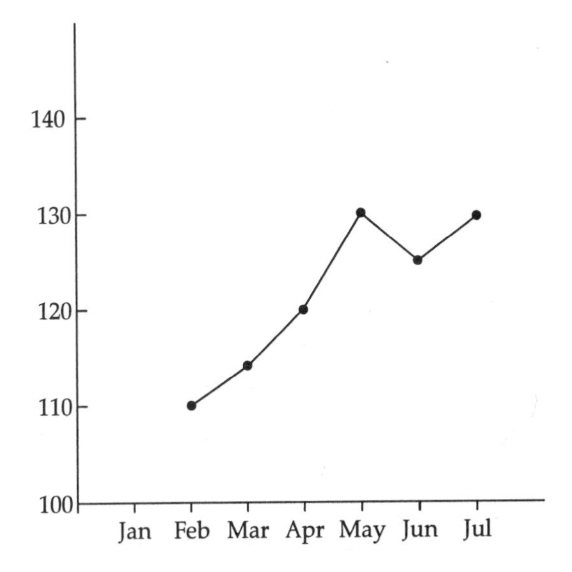

Figure 8.6 Graph of moving averages
Note that the average is located at the middle of the time period.

The forecasting of sales demand and market trends are the most difficult in which to gain accuracy because of the lack of certainty in the data. However, having produced a sales forecast the forecasting of most of the internal requirements of the business, is more certain because of the availability of proven data about internal operations. For example, once a sales demand of 100 units has been decided, then Manufacturing can forecast labour requirements from the standard hours needed to produce one unit. Also Purchasing and Inventory Control/MRP can forecast the materials required to be purchased and stocked from the parts list or Bill of Materials for the one unit. From these calculations Finance can forecast the cash flows needed to finance the activities by applying standard costs and other financial criteria.

8.4 Exercise 8

1. What are the two major types of budget and how do they help management in the running of a business?

2. Draw a budget structure to include the following:

- Master budget
- Design department
- Production department
- Sales department
- Cash budget
- Materials budget.

3. A production department has the following data:

Output	1000 per months 1 and 2
Output	1200 month 3
Labour hours	2 hours per unit
Labour cost	£5.00 per hour
Indirect labour overhead	10% of direct labour cost
Direct materials	£1.50 per unit
Material overhead	10% of direct material cost
Variable overheads	5% of direct labour
Fixed overheads	£300 per month

Construct an operating budget for three months showing the total labour, materials and overhead figures plus the grand total for each month.

4. Detail four stages of budgeting and describe their major functions.

5. How are budgets normally divided up into time periods, and what are the advantages of doing this?

6. (a) What is the purpose of cost centre codes? (b) Construct a typical three-tier code covering two engineering functions, devising your own cost centre codes.

7. What is meant by inventory? Name four types of material that might be carried in an engineering company's inventory.

8. What are stock and work in progress?

9. Name three elements of inventory cost.

10. What do the terms FIFO and LIFO stand for? What is different with these in the way that stock is valued?

11. (a) Name three commonly used periods of forecasting. (b) Describe four items or areas of activity that might be forecast.

12. Construct a 12-month forecast of sales from the following data:

Product A

Month 1 output	1000 units
Months 2 to 6	increase on month 1 by 2%
Month 7	700 units
Month 8	1100
Months 9 to 12	increase by 2% cumulative on month 8

Product B

Month 1	500 units
Months 2 to 12	increase by 3% per month cumulative (use standard rounding of 0.5+)

Product C

Start production in month 5

Months 5 to 8	200 per month
Months 9 to 12	10% cumulative decrease on month 8

Show the grand total for all three products in each month.

13. Draw a stock level graph for three months (13 weeks) based on the following data:

Opening stock	200
Consumption	50 per week
Buffer stock held	1 week consumption
Order lead time	2 weeks
Order interval	4 weeks

Plot the stock levels and identify the re-order points in terms of the week numbers.

14. Using the total of units from question 12, draw a simple trend graph, and moving average graphs for the calendar year. Assume totals of 1200 in month 12 of the prior year and that the figures for month 1 of the next year follow the same calculation as month 12.

Appendix 1

Present value factor for discounted cash flow calculations

Table A1 Present value factors for discounted cash flow calculations

After year	1%	2%	3%	4%	5%	6%	7%	8%	9%	10%	11%	12%	13%	14%	15%
1	0.9901	0.9804	0.9709	0.9615	0.9524	0.9434	0.9346	0.9259	0.9174	0.9091	0.9009	0.8929	0.8850	0.8772	0.8696
2	0.9803	0.9612	0.9426	0.9246	0.9070	0.8900	0.8734	0.8573	0.8417	0.8264	0.8116	0.7972	0.7831	0.7695	0.7561
3	0.9706	0.9423	0.9151	0.8890	0.8638	0.8396	0.8163	0.7938	0.7722	0.7513	0.7312	0.7118	0.6931	0.6750	0.6575
4	0.9610	0.9238	0.8885	0.8548	0.8227	0.7921	0.7629	0.7350	0.7084	0.6830	0.6587	0.6355	0.6133	0.5921	0.5718
5	0.9515	0.9057	0.8626	0.8219	0.7835	0.7473	0.7130	0.6806	0.6499	0.6209	0.5935	0.5674	0.5428	0.5194	0.4972
6	0.9420	0.8880	0.8375	0.7903	0.7462	0.7050	0.6663	0.6302	0.5963	0.5645	0.5346	0.5066	0.4803	0.4556	0.4323
7	0.9327	0.8706	0.8131	0.7599	0.7107	0.6651	0.6227	0.5835	0.5470	0.5132	0.4817	0.4523	0.4251	0.3996	0.3759
8	0.9235	0.8535	0.7894	0.7307	0.6768	0.6274	0.5820	0.5403	0.5019	0.4665	0.4339	0.4039	0.3762	0.3506	0.3269
9	0.9143	0.8368	0.7664	0.7026	0.6446	0.5919	0.5439	0.5002	0.4604	0.4241	0.3909	0.3606	0.3329	0.3075	0.2843
10	0.9053	0.8203	0.7441	0.6756	0.6139	0.5584	0.5083	0.4632	0.4224	0.3855	0.3522	0.3220	0.2946	0.2697	0.2472
11	0.8963	0.8043	0.7224	0.6496	0.5847	0.5268	0.4751	0.4289	0.3875	0.3505	0.3173	0.2875	0.2607	0.2366	0.2149
12	0.8874	0.7885	0.7014	0.6246	0.5568	0.4970	0.4440	0.3971	0.3555	0.3186	0.2858	0.2567	0.2307	0.2076	0.1869
13	0.8787	0.7730	0.6810	0.6006	0.5303	0.4688	0.4150	0.3677	0.3262	0.2897	0.2575	0.2292	0.2042	0.1821	0.1625
14	0.8700	0.7579	0.6611	0.5775	0.5051	0.4423	0.3878	0.3405	0.2992	0.2633	0.2320	0.2046	0.1807	0.1597	0.1413
15	0.8613	0.7430	0.6419	0.5553	0.4810	0.4173	0.3624	0.3152	0.2745	0.2394	0.2090	0.1827	0.1599	0.1401	0.1229
16	0.8528	0.7284	0.6232	0.5339	0.4581	0.3936	0.3387	0.2919	0.2519	0.2176	0.1883	0.1631	0.1415	0.1229	0.1069
17	0.8444	0.7142	0.6050	0.5134	0.4363	0.3714	0.3166	0.2703	0.2311	0.1978	0.1696	0.1456	0.1252	0.1078	0.0929
18	0.8360	0.7002	0.5874	0.4936	0.4155	0.3503	0.2959	0.2502	0.2120	0.1799	0.1528	0.1300	0.1108	0.0946	0.0808
19	0.8277	0.6864	0.5703	0.4746	0.3957	0.3305	0.2765	0.2317	0.1945	0.1635	0.1377	0.1161	0.0981	0.0829	0.0703
20	0.8195	0.6730	0.5537	0.4564	0.3769	0.3118	0.2584	0.2145	0.1784	0.1486	0.1240	0.1037	0.0868	0.0728	0.0611
21	0.8114	0.6598	0.5375	0.4388	0.3589	0.2942	0.2415	0.1987	0.1637	0.1351	0.1117	0.0926	0.0768	0.0638	0.0531
22	0.8034	0.6468	0.5219	0.4220	0.3418	0.2775	0.2257	0.1839	0.1502	0.1228	0.1007	0.0826	0.0680	0.0560	0.0462
23	0.7954	0.6342	0.5067	0.4057	0.3256	0.2618	0.2109	0.1703	0.1378	0.1117	0.0907	0.0738	0.0601	0.0491	0.0402
24	0.7876	0.6217	0.4919	0.3901	0.3101	0.2470	0.1971	0.1577	0.1264	0.1015	0.0817	0.0659	0.0532	0.0431	0.0349
25	0.7798	0.6095	0.4776	0.3751	0.2953	0.2330	0.1842	0.1460	0.1160	0.0923	0.0736	0.0588	0.0471	0.0378	0.0304

Table A1 continued

After year	16%	17%	18%	19%	20%	21%	22%	23%	24%	25%	26%	27%	28%	29%	30%
1	0.8621	0.8547	0.8475	0.8403	0.8333	0.8264	0.8197	0.8130	0.8065	0.8000	0.7937	0.7874	0.7813	0.7752	0.7692
2	0.7432	0.7305	0.7182	0.7062	0.6944	0.6830	0.6719	0.6610	0.6504	0.6400	0.6299	0.6200	0.6104	0.6009	0.5917
3	0.6407	0.6244	0.6086	0.5934	0.5787	0.5645	0.5507	0.5374	0.5245	0.5120	0.4999	0.4882	0.4768	0.4658	0.4552
4	0.5523	0.5337	0.5158	0.4987	0.4823	0.4665	0.4514	0.4369	0.4230	0.4096	0.3968	0.3844	0.3725	0.3611	0.3501
5	0.4761	0.4561	0.4371	0.4190	0.4019	0.3855	0.3700	0.3552	0.3411	0.3277	0.3149	0.3027	0.2910	0.2799	0.2693
6	0.4104	0.3898	0.3704	0.3521	0.3349	0.3186	0.3033	0.2888	0.2751	0.2621	0.2499	0.2383	0.2274	0.2170	0.2072
7	0.3538	0.3332	0.3139	0.2959	0.2791	0.2633	0.2486	0.2348	0.2218	0.2097	0.1983	0.1877	0.1776	0.1682	0.1594
8	0.3050	0.2848	0.2660	0.2487	0.2326	0.2176	0.2038	0.1909	0.1789	0.1678	0.1574	0.1478	0.1388	0.1304	0.1226
9	0.2630	0.2434	0.2255	0.2090	0.1938	0.1799	0.1670	0.1552	0.1443	0.1342	0.1249	0.1164	0.1084	0.1011	0.0943
10	0.2267	0.2080	0.1911	0.1756	0.1615	0.1486	0.1369	0.1262	0.1164	0.1074	0.0992	0.0916	0.0847	0.0784	0.0725
11	0.1954	0.1778	0.1619	0.1476	0.1346	0.1228	0.1122	0.1026	0.0938	0.0859	0.0787	0.0721	0.0662	0.0607	0.0558
12	0.1685	0.1520	0.1372	0.1240	0.1122	0.1015	0.0920	0.0834	0.0757	0.0687	0.0625	0.0568	0.0517	0.0471	0.0429
13	0.1452	0.1299	0.1163	0.1042	0.0935	0.0839	0.0754	0.0678	0.0610	0.0550	0.0496	0.0447	0.0404	0.0365	0.0330
14	0.1252	0.1110	0.0985	0.0876	0.0779	0.0693	0.0618	0.0551	0.0492	0.0440	0.0393	0.0352	0.0316	0.0283	0.0253
15	0.1079	0.0949	0.0835	0.0736	0.0649	0.0573	0.0507	0.0448	0.0397	0.0352	0.0312	0.0277	0.0247	0.0219	0.0195
16	0.0930	0.0811	0.0708	0.0618	0.0541	0.0474	0.0415	0.0364	0.0320	0.0281	0.0248	0.0218	0.0193	0.0170	0.0150
17	0.0802	0.0693	0.0600	0.0520	0.0451	0.0391	0.0340	0.0296	0.0258	0.0225	0.0197	0.0172	0.0150	0.0132	0.0116
18	0.0691	0.0592	0.0508	0.0437	0.0376	0.0323	0.0279	0.0241	0.0208	0.0180	0.0156	0.0135	0.0118	0.0102	0.0089
19	0.0596	0.0506	0.0431	0.0367	0.0313	0.0267	0.0229	0.0196	0.0168	0.0144	0.0124	0.0107	0.0092	0.0079	0.0068
20	0.0514	0.0433	0.0365	0.0308	0.0261	0.0221	0.0187	0.0159	0.0135	0.0115	0.0098	0.0084	0.0072	0.0061	0.0053
21	0.0443	0.0370	0.0309	0.0259	0.0217	0.0183	0.0154	0.0129	0.0109	0.0092	0.0078	0.0066	0.0056	0.0048	0.0040
22	0.0382	0.0316	0.0262	0.0218	0.0181	0.0151	0.0126	0.0105	0.0088	0.0074	0.0062	0.0052	0.0044	0.0037	0.0031
23	0.0329	0.0270	0.0222	0.0183	0.0151	0.0125	0.0103	0.0086	0.0071	0.0059	0.0049	0.0041	0.0034	0.0029	0.0024
24	0.0284	0.0231	0.0188	0.0154	0.0126	0.0103	0.0085	0.0070	0.0057	0.0047	0.0039	0.0032	0.0027	0.0022	0.0018
25	0.0245	0.0197	0.0160	0.0129	0.0105	0.0085	0.0069	0.0057	0.0046	0.0038	0.0031	0.0025	0.0021	0.0017	0.0014

Appendix 2

Elements of Labour Costs

The total labour or 'employment costs' will include direct costs and overhead costs.

Direct costs (employee renumeration)

- Basic salary
- Shift allowance
- Overtime payments
- Incentive payments

Overhead costs

- Sickness, holiday and maternity pay
- Employer's pension contribution
- Employer's National Insurance contribution

1 Direct costs

Basic salary

Shop floor staff – pay is usually calculated on an hourly or weekly rate. In the past this has normally been paid weekly in cash on a Thursday or Friday for the preceding week's work. An increasing number are being paid monthly as company payroll arrangements move towards an all monthly pay system which automatically lodges the net amount due into the employee's bank account or building society account. This trend is accelerating as modern personnel harmonisation programmes are introduced to remove traditional demarcations between shop floor and office staff and equalise the conditions of employment.

Office staff – junior clerical posts may be weekly paid but many junior specialist and all senior posts are given an annual salary paid in equal monthly amounts. This is usually paid directly into the employee's chosen personal bank account.

Shift allowances

These vary greatly with the type of shift working. Two commonly seen types are continuous and permanent shifts. Continuous 24-hour working in 2 or 3 shifts where employees work for a fixed period each shift in turn, changing shifts on a weekly or other number of days basis. These usually provide a constant rate of pay regardless of which shift or shifts are worked within a payment period.

Permanent shifts of 8, 10 or 12 hours that do not rotate employees. These shifts will normally pay differing rates. The permanent day shift will receive a basic wage at levels normal to the area or industry. The night shift and other shifts will normally receive a premium of between 25% and 33% of the basic rate.

Overtime

Overtime is normally calculated as a premium on the basic salary. Typical premium rates for staff employed Monday – Friday are:

- 25% for evening or early morning work
- 33% for Saturday morning
- 50% for Saturday afternoon
- 100% for Sunday

Again, premium rates will vary with company, site and industry.

Incentive payments and bonuses

The emphasis is moving away from schemes based on the measurement of individual output as group working aimed at team work and higher quality levels is introduced. Where individual or group incentives are used, they are usually paid in relation to the normal payment period, i.e. weekly or monthly. Alternatively, some companies pay a bonus which is based on the annual financial results of the company or operating division. This goes to all employees and may be a fixed sum or a percentage of annual salary.

2 Overhead costs

Sickness and maternity pay

These are normally calculated as a proportion of salary and are part of the company's standard working conditions. There are also statutory requirements in force through the application of the Statutory Sick Pay (SSP) and Statutory Maternity Pay schemes of the DHSS.

Employer's pension contributions

Most medium and large companies have a company pension scheme. In these cases the employer will make a contribution of between 5 and 10 per cent of the employee's basic + shift premium wage. Most schemes also require a contribution from the employee of typically between 3 and 6 per cent of salary. A smaller proportion of schemes are non-contributory. The majority are 'Final Salary' schemes which provide a pension based on a fraction of the employee's final salary (or an average of some final years) times the length of service e.g. final salary × 1/60 for each year of service. In many schemes a maximum pension of two-thirds of the final salary is payable. Thus, in a 1/60 scheme, 40 years' service is needed to provide a 40/60 or 2/3 final salary as a maximum annual pension.

The question of pensions is becoming a major concern for companies and employees as the number and longevity of pensioners increase in relation to the number of employed persons thus creating a larger overhead burden for the nation. The government has already indicated that it will be reducing state pensions in the future because of this increasing burden on the state.

Also it is becoming less common for staff to stay at one company for the number of years needed to secure a 50 or 66 per cent pension from a final salary scheme. Recent years have seen the development of Executive, Individual and Personal pension schemes which acrue contributions to provide a lump sum at the end. This is the property of and is invested by the pensioner in a number of alternative private financial arrangements. The employer will contribute to Executive or Individual schemes as part of the renumeration package.

Employer's National Insurance contributions

National Insurance has to be paid by both the employer and the employee for all employees who earn above a minimum figure for a week or month. The employer and employee make contributions calculated as a percentage of the salary, the percentage applicable being on a sliding scale to an upper limit for the employee but not the employer. The amount depends on whether the company pension scheme is 'contracted in' or 'contracted out' of the government pension earnings related scheme. In most cases an employer will pay about 7 per cent of salary if contracted out and 10 per cent of salary if contracted in. The employee will pay up to 10 per cent on a sliding scale.

3 Summary

National Insurance, pension, holiday, sick pay and maternity pay contributions represent an employment overhead of typically 20 per cent of the employee's basic pay. This amounts to £1000 – £6000 per annum for persons within the current range of salaries normally paid and must be taken into account when considering:

- internal manpower planning and budgeting;
- comparing in-house costs with sub-contract or purchase;
- the justification of capital expenditure aimed a reducing internal labour content.

Appendix 3

Elements of Material Costs

Materials are consumed in a number of ways and areas within a business. This appendix outlines some of these.

1 Production materials

Repetitive production

The greatest company expenditure on materials will normally be for production materials that go into the product. This material is regarded as stock and will appear in various guises as it moves through the production cycle, e.g. raw materials, purchased items, work-in-progress, kanban stock or finished goods. It appears on the inventory budgets of Production and Stores departments but rarely in Engineering budgets.

Low-volume capital goods production

Here, a sale is often handled as a project throughout the design and manufacturing cycle. In this case a Project Manager or Project Engineer may well have responsibility for the final cost of the project including materials and labour costs. Inventory control may still be operated through the factory MRP or stock control system but the material costs will also be booked to each project.

2 Engineering materials

Engineering Departments

With the exception of capital projects (see below), engineering departmental expenditure on materials will be low compared to their labour costs. Departments such as Research & Development laboratories, Toolrooms and Maintenance will be the largest spenders. Office departments such as Design/DO, Manufacturing Engineering and QA should have very low materials costs relative to labour costs. The engineering materials unit costs are likely to be high in comparison to production materials because the former may be specials or purchased in low quantities.

Engineering Capital Projects

Such projects that are off-site will need a method of controlling materials and their cost. Materials tracking may take various forms of booking and stock control but MRP would not be generally used in this one-off situation. Material costs can be tracked and analysed on a project management package using PERT techniques. An engineer may be responsible for materials control and costs either as Project Manager/Engineer or Clerk of Works in civil engineering.

Consumable materials and items

These are generally classified as:

1. General bulk materials consumed in production that it is difficult to apportion to the direct cost of any particular product type. Examples are oils, adhesives, welding rods, plating process materials, paint, etc.
2. Similarly, materials used by the support departments including engineering. Examples are stationery, drawing paper, writing implements, floppy disks, and tapes. The level of these cost will be low when compared with the cost of staff but nevertheless they need to be kept under control as they are items that are suitable for personal use at home.

Recommended further reading

Balkwell, J. and Freeman-Bell, G. (1964) *Management in Engineering,* London: Prentice Hall.

Batty, J. (1969) *Industrial Administration and Management,* London: MacDonald and Evans.

Evans, E.C. (1969) *The Pillars of Management Accounting Series,* vols 1, 2 and 3, London: MacDonald.

Henderson, S. *et al.* (1993) *Management for Engineers,* Oxford: Butterworth Heinemann.

Institution of Production Engineers and Institution of Costs and Works Accountants (1969) *An Engineer's Guide to Costing,* London: I.Prod.E and ICWA.

Koshal, D. (ed.) (1993) *Manufacturing Engineers Reference Book,* Oxford: Butterworth Heinemann.

Lock, D. (ed.) (1992) *Handbook of Engineering Management,* Oxford: Butterworth Heinemann.

Rockley, L.E. (1970) *Finance for the Non-Accountant,* London: Business Books.

Rockley, L.E. (1973) *The Non-Accountant's Guide to the Balance Sheet,* London: Business Books.

Answers to Exercises

Note: Some of the answers can only be specimen answers as they are intended to start the student thinking about the particular issue in his or her organisation.

Exercise 1

1. Details of student's organisation needed, e.g. electrical manufacturer = secondary sector; mining company = primary sector.
2. Sample answer
   - primary – mining, agriculture, forestry, fishing
   - secondary – manufacturing, steel production, construction
   - tertiary – banking, insurance, retail, transport.
3. The limited company is the most common, because it limits the liability of its shareholders to the value of shares that they own.
4. Private Limited Company is the most likely option. The options are:
   - Private Limited Company
   - Partnership
   - Self-employed
   - Sole trader.
5. This restricts the liability and potential losses of a shareholder to the price that was paid for the shares.
6. See the list in section 1.3 The Business Cycle.
7. Three factors are likely to be:
   - whether a standard design is used or if a product is designed to order.
   - whether the business will supply to customers against credit or not. This will be set by the selling policy.
   - whether the business can obtain credit from its suppliers.
8. Groups of similar work:
   - Design – Design and specify the product.
   - Commercial – Obtain orders.
     – Purchase materials.

• Manufacturing	– Make the product. – Test the product. – Deliver the product.
• Financial	– Invoice the customer. – Receive payment. – Pay for materials, services. – Pay wages.

Exercise 2

1. (a) Sales
   Purchasing
   Financial
   (b) Sales
   Purchasing
   Financial
   Production
   (c) Design
   Production
   Purchasing
   Sales
   Financial
   (d) Marketing
   Sales
   Research and Development
   Design
   Manufacturing
   Quality Assurance
   Purchasing
   Finance
   Personnel
2. The other two major functions are:
   - Product Design
   - Manufacturing.

   Examples of primary tasks are:
   (a) Product Design:
   - Research and Development
   - Design Office
   - Application Engineering/Sales Support.

   (b) Manufacturing
   - Production Department
   - Production Control
   - Manufacturing Engineering.

3. **Table A1** Functions in operation

	Sub-contract machining	Design and manufacture of washing machines	Design of printed circuit boards
R and D		X	X
Product design		X	X
Purchasing	X	X	X
Manufacturing	X	X	
Man. engineering	X	X	
Sales	X	X	X
After-sales Support		X	X
Finance	X	X	X

4. **Table A2** Appropriate levels of managerial personnel

Staff	Levels
10	1
50	2
500	4
1000	5

5. The managerial organisation structure is designed to transmit decisions and information up and down the line of managerial command from Senior Managers to supervisors and vice versa. However, the product and therefore important information about it, flows across the organisation from Sales order design manufacture to despatch, needing inter-departmental communication.
6. (a) Flat organisation: an organisation with one or very few levels of supervision and management.
   (b) Hierarchical: where there are a number of management levels reporting up a chain of command and usually organised by function e.g., Sales, Manufacturing.
   (c) Matrix: an organisation that combines hierarchical levels of management with functional specialisation. Staff from different functions may be combined to form a project team and report to the Project Manager as well as their core function manager.
7. (a) Businesses of 20 staff producing one product is best suited by a flat organisation because the number of staff to control are small and they are all likely to be familiar with the single product.
   (b) A business of 500 people with a wide range of products could be either hierarchical or matrix. The modern trend is for matrix management so that teams from the different functions can identify with a particular product, product group or customer.

8. A centralised organisation is one where decision-making is centred on one manager or place and the whole organisation is controlled from that centre. A decentralised organisation is one where decision-making is spread to separate functions or locations and local managers have a greater say in the running of their local activities.
9. (a) Supervising is concerned with the guidance and control of people on a daily basis. Supervision is one of the activities common to supervisors, managers and directors.
   (b) Deploying is the allocation of resources and people to particular tasks.
10. Controlling activities are those which monitor, measure and direct action against targets or aims. Controlling is used to feed back information to the planning activity which sets out the programmes of work and targets in the first place.
11. Facilitating is an activity that assists persons to achieve targets or improvements by providing resources or information to help them do so. Co-ordinating is concerned with the correct matching of activities particularly in relation to a time schedule. It is making sure that the right thing happens at the right time.
12. Line management is the chain of management and managers, within the functions of a typical hierarchical organisation. Line management supervisors in production report to managers in production who report to the Production Director. See Figure 2.9. Staff management often provides a service and specialist input to the line management. See Figure 2.10.

Exercise 3

1. (a) Product Design and Sales Support.
   This encompasses the product technology and knowledge that the company is selling to its customers. Research and Development Engineers, Product Design Engineers and Product Support Engineers from Sales all contribute to the fund of knowledge that goes into the product and its use in service.
   (b) Manufacturing Process Support (Manufacturing Engineering).
   This body of knowledge deals with the manufacturing processes and the production of the product. Some processes are particular to certain products, for example in textiles, oil, chemicals and electronics, whilst other processes such as general machining and welding are applicable to many products.
   (c) Site facilities (Facilities or Plant Engineering).
   This is concerned with providing general engineering level services to all the facilities used on a site, for example, power, light, heating, drainage, buildings, etc.
   (d) Quality Assurance and Support.
   This ensures that the quality of the product produced by the processes meets the quality standards specified by the product

design. In this sense it is a link between Product Design and Manufacturing Engineering. Quality Assurance in its widest sense also ensures that the design and manufacturing control systems employed operate to certain standards, e.g., ISO 9000 series.

2. No, they are not. Engineers are often employed in an area that uses their expertise but is not necessarily an area dealing solely with their type of discipline.

Example: An electrical engineer may work in:

- Product Design Department on designing electrical products, or
- Purchasing Department on purchasing electrical components, or
- Site Engineering maintaining the power distribution network for the site.

3. Refer to page 36 – Product Design and Sales Support.
4. Refer to page 39 – Manufacturing Process Design and Support.
5. Refer to page 45 – Quality Assurance and support.
6. Refer to page 43 – Facilities Engineering.
7. Manufacturing Engineering and Facilities Engineering. Because both are concerned with providing production resources and the technologies may overlap. If so, these can be handled by one engineer or department covering both aspects.
8. (a) R & D will investigate the possibilities of new products, features or materials but will not normally design a new product in full for production. They will pass a product outline or research findings to Product Design.
   (b) Product Design will design a product in full and produce a performance specification for the product. These will then be passed to the Drawing Office.
   (c) Drawing Office will undertake the detailed drawing and specification of all the parts and materials required to manufacture the product.
9. (a) Process Engineering applies expertise in particular production processes, for example, CNC turning, welding or printed circuit board manufacture. It will normally be the department that specifies process equipment and its tooling.
   (b) Industrial Engineering is usually associated with the organisation of the workplace, process times and process costs. Industrial Engineering will often provide costs and time estimates for the accounts and production departments.
   (c) Line Engineering provides resident manufacturing engineers to troubleshoot technical problems on a production line or for a given product group. This may involve elements of Process Engineering and Industrial Engineering. Line Engineers will feed back production engineering problems for longer-term solution by the process or production engineers.

10 (a) Quality Assurance is concerned with the correct functioning of systems that control the quality of the company's activities. This includes commercial departments as well as engineering and production.

(b) Quality Control deals directly with the quality of production and the materials used. It normally incorporates the various inspection functions that monitor the quality of incoming materials, production output and finished product testing.

(c) Quality Engineering deals with the metrology used in manufacturing. It will specify, devise and monitor inspection equipment for use by Quality Control and Production to ensure that the inspection equipment itself follows national and international standards.

11.

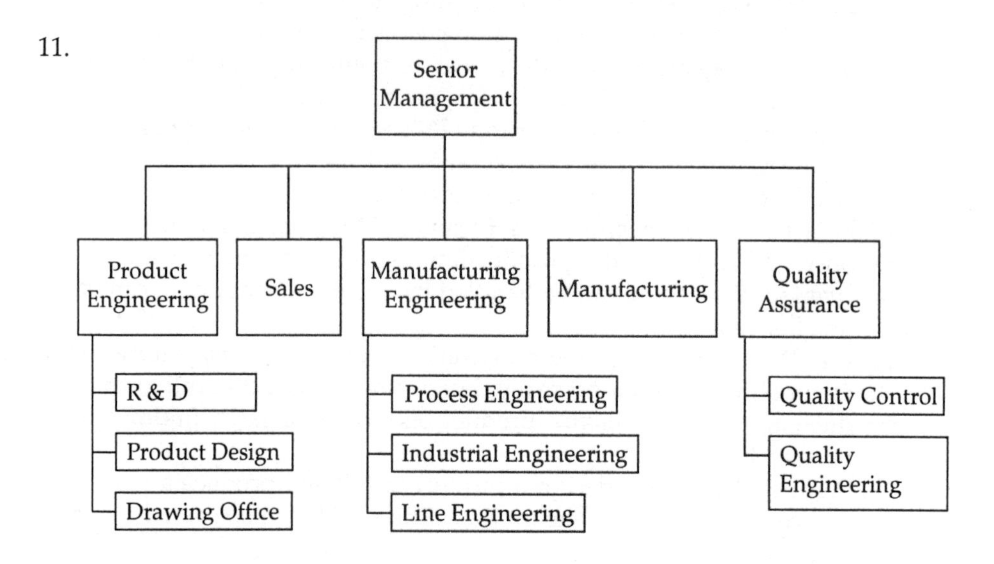

Figure A1 Organisation chart

12. **Table A3** Types of production and industries

Type	Product examples
one-off	Where products are made to an order and/or special one-off design: e.g. shipbuilding civil engineering special machines
batch	Where a number of standard products are made the same on the same production equipment: e.g. electric motors plastic mouldings aircraft
continuous	Where continuous high-value production of the same product is needed: e.g. dry batteries breweries volume car production

13. **Table A4** Modes of production

Mode	Product examples
dedicated lines	high-volume short production cycle: e.g. food and drink volume cars toiletries
process centred	products needing predominantly one type of production process: e.g. sub-contract machining engineering products plastic mouldings
cell manufacturing	Any product made in medium size quantities with some variety of model and processes involved
Just in Time	Any product that is not high volume and short production cycle and would justify a dedicated line

14. Cell production reduces total lead time, travelling time between processes and improves motivation of staff.

Exercise 4

1. (a) Market Research to find out what the market wants.
   (b) Product targeting to a particular market.
   (c) Product promotion in the market.
2. (a) As a separate function from Sales and reporting directly to the Chief Executive.
   (b) As a combined function with Sales in which the manager would report to a Marketing and Sales Executive.
3. The essential difference is that marketing is concerned with the placement and promotion of the product in the general market. Sales is concerned with directly selling to customers and supporting the customer's needs.
4. (a) Least risk: increase market share of existing products in existing markets – lowest costs and greatest certainty of success.
   (b) Highest risk: develop new products for new markets – highest cost and greatest uncertainty of success.
5. To provide market information on what might be the most promising line of enquiry in researching new products, materials and processes.
6. Key objectives of Sales Function:
   (a) Organise sales activities within the markets.
   (b) Produce forecast sales – usually for a year.
   (c) Recruit and train sales staff.
   (d) Establish and maintain good relations with existing and potential customers.

7. Sales function can be organised by product type or sales area. The advantage of organisation by product type is that greater expertise can be applied by specialisation, particularly with technical products.
8. To give the organisation a sales plan that will be the basis for the planning of all other activities.
9. Very important in all cases.
10.

Average units	Q1+Q2	229 175
Q3 units	–2850	–9.7%
Q4 units	+4725	+17.7%
Average value	Q1+Q2	£22 1250
Q3 value	–£20 250	–9.1%
Q4 value	+£39 250	+17.7%

11. Domestic heaters, gas, electricity, Christmas cards, toys, clothing, shoes, sports equipment, gardening equipment, holiday hotels, rail and air travel.
12. (a) Product Engineering can support Sales by:
    (i) Providing technical information and specifications on the product for sales literature and advertising.
    (ii) Providing technical information in answer to customers or sales enquiries on product models to be used.
    (b) The Quality Assurance function supports Sales by:
    (i) ensuring that the quality of the product delivered is correct.
    (ii) investigate and distribute information on customer complaints and faulty products in service.
13. Marketing and Sales function or Operations function.
14. High volume consumer goods. For example, products in the food, drink and clothing industries.
15. Design and layout of warehouse racking and handling equipment.
16. The recording and tracking of rejected products from the customer or goods damaged in transit/warehousing.
17. Key functions e.g.:
    (a) Sending of technical specification and delivery requirements to potential customers.
    (b) Obtaining quotations.
    (c) Selecting suppliers.
    (d) Placing orders.
    (e) Controlling progress of orders.
18. Contract agreement and separate orders.
19. Materials Requirements Planning (MRP).
20. Four factors from twelve listed on page 75.
    e.g. quality guarantees
    quantity delivered
    price
    previous record as a supplier.
21. It is important that the buyer's commercial and contractual knowledge is brought to bear on negotiations in order to avoid disadvantage to the

organisation. With technical goods, engineers are often heavily involved in the selection but may agree to unsuitable commercial conditions.

Exercise 5

1. Two main branches are financial accounting and management accounting. Financial accounting deals with the overall financial position of the company or business and prepares the accounts. Management accounting deals with the financial control of departments and their costs.
2. See Figure 5.2.
3. Balance Sheet and Profit and Loss Account.
4. To present an overview or snapshot of the assets and liabilities of an organisation at a particular time. It must balance the debits and credits. Usually valid for the last day of the trading period under review.
5. Refer to figure 5.3 to create balance of £100 000.
6. The profit and loss account summarises the trading position and the income and expenditures in the period under review.
7. The three cost categories are material costs, labour costs and overhead costs.
8. Capital expenditure is that spent on fixed assets such as buildings and production equipment. Operational expenditure is used to buy consumable items, materials and services including labour.
9. See list of categories on page 87.
10. Data needed
    - cost of equipment
    - cost to install and service
    - cost of each item produced on machine
    - cost of item by existing method
    - savings per item
    - annual production volume which will give annual savings.
11. Three methods used in investment approval are:
    - Payback
    - Net Present Value (NPV)
    - Internal Rate of Return (IRR).

    The payback method will only measure the time taken to recover the cost. It will not indicate the profitability. It is, however, very simple to understand as a non-financial indicator.
12. (a) Net present value £38 539
    (b) Internal rate of return £16%
13. Cost to make £13.225
    Cost to buy £11.742
    Saving by buying £ 2.501
    Decision should be to buy the part.

14. **Table A5** Profitability calculations

Sales level (units)	**100** £	**200** £	**300** £	**400** £	**500** £	**132**	**423**
Revenue @ £10.00 each	1000	2000	3000	4000	5000	1320	4335.75
Variable Costs							
Cost per unit	5.2	5.2	5.2	5.2	5.2	5.2	5.2
Total variable cost	520	1040	1560	2080	2600	686.4	2199.6
Semi-variable cost	100	400	600	800	1000	132	634.5
Total-variable costs	620	1440	2160	2880	3600	818.4	2834.1
Contribution	380	560	840	1120	1400	501.6	
Fixed cost	500	500	500	500	500	500	1500
Fixed + variable cost	1020	1540	2060	2580	3100	1186.4	3699.6
Grand total cost	1120	1940	2660	3380	4100	1318.4	4334.1
Profit	–120	60	340	620	900	1.6	

Exercise 6

1. Note: specimen answers of the more important interfaces.
   (a) Research and Development
      Product Design, Marketing, Research Laboratories, Manufacturing Engineering, Universities.
   (b) Product Design
      R & D, Drawing Office, Manufacturing Engineering, Quality Control, Materials Suppliers.
   (c) Drawing Office
      Product Design, Manufacturing Engineering, Quality Control, Production.
   (d) Manufacturing Engineering
      Production Departments, Product Design, Drawing Office, Equipment Suppliers.
   (e) Facilities Engineering
      Production, Manufacturing Engineering, outside service suppliers (gas, electricity), outside authorities (fire, Health and Safety).
   (f) Quality Assurance
      Financial Managers, Production Departments, Drawing Office, Manufacturing Engineering.
   (g) Manufacturing
      Drawing Office, Purchasing, Quality Control, Manufacturing Engineering.
   (h) Marketing
      The markets, Sales, R&D, Manufacturing.

(i) Sales
The customers, Marketing, Production, Distribution.
(j) Distribution
Marketing, Sales, Manufacturing, Export/Input Agencies.
(k) Purchasing
Suppliers, Drawing Office, Stores/Inventory Control. Goods in inspection.

2. (a) CAD – Computer Aided Design
   CAE – Computer Aided Engineering
   (b) MRP – Materials Requirements Planning
   SFDC – Shop Floor Data Capture
   (c) CAM – Computer Aided Manufacture
   CAPP – Computer Aided Process Planning

Exercise 7

1. (a) To know how much individual activities have cost.
   (b) To know the true cost of each product.
   (c) To know the true total cost of the business.
2. (a) Elements of cost.
   (b) Type of cost.
   (c) Behaviour of costs.
   (d) Techniques of costing.
3. (a) Labour cost: identifies the cost of labour expended on the operation/item.
   (b) Material cost: identifies the cost of material comprising the item.
   (c) Expense cost: identifies incidental and overhead costs consumed in producing an item.
4. Typically, rent, heating, gas, water, stationery, insurance, depreciation, tooling.
5. Direct costs can be directly attributed to the production of a single product or service. Indirect costs are general costs that are shared by a number of products or services. They are usually included in the cost of production as an additional percentage of the direct cost.
6. (a) Total cost £2.95
   (b) Percentage attributable to labour 53.2%
   (c) Percentage attributable to material 46.8%
7. (a) Cost of operation £2.88
   (b) Total cost after two operations £5.83
8. Prime cost £4.91
   Rent and rates £0.049
   Service overhead £2.201
   Ex-factory cost £7.168
9. Total cost £11.468
   Selling price £14.43

10. Variable costs: labour, materials. Fixed costs: rent, rates.
11. Refer to Figure 5.7 break-even chart as an example.
12. Absorption costing spreads all of the overhead and indirect costs of a production facility over the total units produced in a period.
13. Monthly overhead budget £956
14. Monthly budget Dept A £350
    Monthly budget Dept B £291
    Monthly budget Dept C £175
15. Monthly budget Dept A £358
    Monthly budget Dept B £302
    Monthly budget Dept C £191
16. (a) Overhead rate 19.23%
    (b) Monthly overhead Dept A £480
    (c) Monthly overhead Dept B £432
17. Machine hour rate £6.40
    Overhead rate/unit £0.133
18. Marginal costing adds only the cost of additional labour and materials for additional production, assuming that the factory and production overhead costs have already been absorbed by the previous production. Therefore, marginal costing gives a more accurate picture of the actual extra costs involved in additional volumes of production and a truer figure of profitability for the additional output produced.
19. (a) Marginal cost of 500 units £4880
    (b) Contribution £7620

Exercise 8

1. Departmental operating budgets: these help the organisation to control its expenditures on the daily operations of all activities. Capital expendi-

2.

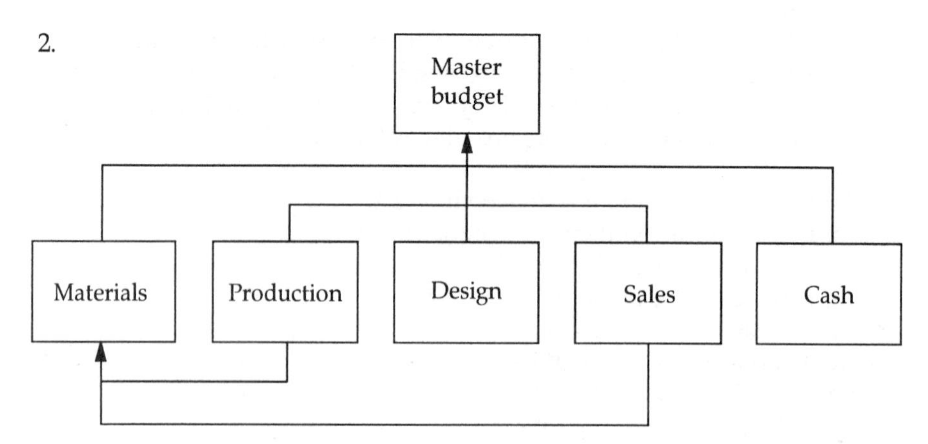

Figure A2 Budget structure

3. **Table A6** Operating budget

Production budget		**Month**	
	1	**2**	**3**
Output (units)	1000	1000	1200
Labour hours @ 2hrs/unit	2000	2000	2400
Direct labour cost @ £5/hr (£)	10000	10000	12000
Labour overhead @ 10% (£)	1000	1000	1200
Total labour budget (£)	11000	11000	13200
Direct material (£)	1500	1500	1800
Material overhead @ 10% (£)	150	150	180
Total material budget (£)	1650	1650	1980
Variable expenses @ 5% Lab (£)	500	500	600
Rent and rates (£)	300	300	300
Total overhead budget (£)	800	800	900
Grand total budget (£)	13450	13450	16080

ture budgets: these help the organisation to control large expenditure and investments in buildings and equipment.

4. Four stages of budgeting:
   (a) Forecasting: to provide a forward plan of likely demands and levels of activity.
   (b) Preparation: the drawing together of all data to produce a budget of all the costs anticipated in a particular period.
   (c) Monitoring: the issue of measured actual costs and their comparison against the budget to establish the variance.
   (d) Adjustment: the activity of amending the budget for the future period to reflect the performance to date.
5. By years, months and weeks. The advantage of this system is to aid the accurate planning of activities, expenditure and incomes over the financial year.
6. (a) Purpose of cost centres is to accurately allot costs to the area of activity that is incurring them (usually a department).
   (b) Student to construct a code structure which should be clearly and logically hierarchical.
7. Inventory includes the total stock of all materials in the organisation. Types of inventory in an engineering company include:
   - raw materials – bar stock and casting
   - purchased components – fastenings, motors, switches
   - components made – output from the production shops
   - finished goods – completed products awaiting despatch

12. **Table A7** Sales forecast and Simple trends

Sales forecast	**Simple trends**													
Month	**12**	**1**	**2**	**3**	**4**	**5**	**6**	**7**	**8**	**9**	**10**	**11**	**12**	**1**
Product														
A		1000	1020	1020	1020	1020	1020	700	1100	1122	1144	1167	1191	1214
B		500	515	530	546	563	580	597	615	633	652	672	692	713
C						200	200	200	200	180	162	146	131	118
Total	1200	1500	1535	1550	1566	1783	1800	1497	1915	1935	1959	1985	2014	2045
Moving average		1412	1528	1551	1633	1716	1693	1737	1782	1936	1960	1986	2015	

8. (a) Stock is materials held in stores waiting to be used.
   (b) Work in progress is materials moving through the production cycle in the production departments.
9. (a) Purchase cost of the materials.
   (b) Administrative cost of purchasing.
   (c) Internal storage costs.
10. First in first out: the cost of the first material in is used in costing.
    Last in first out: the cost of the last material in is used in costing.
11. (a) Long-term, medium-term and short-term forecasting.
    (b) See list on page 158.
12. See page 190.
13.

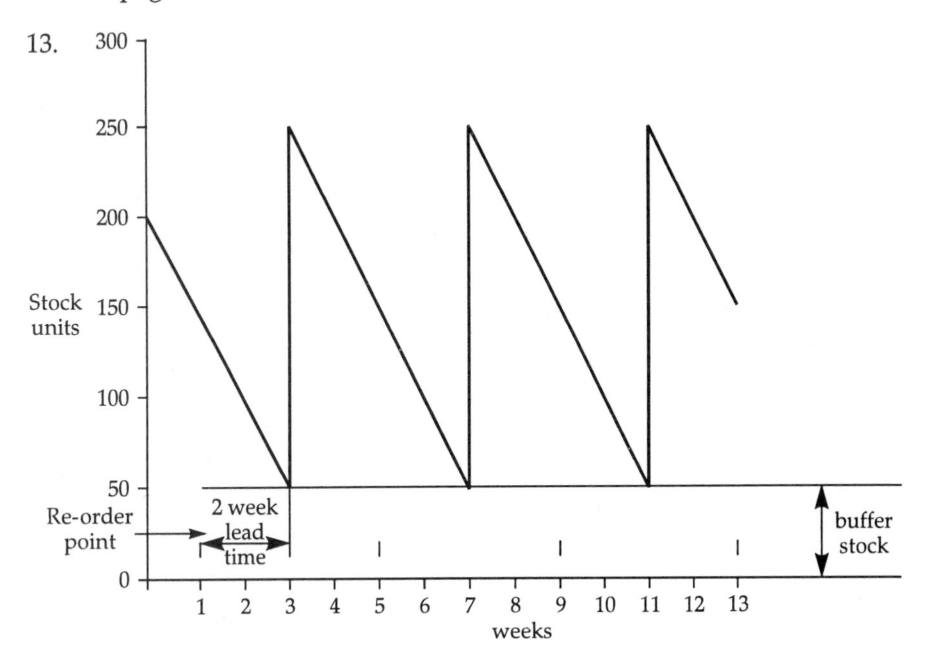

Figure A3 Stock level graph

14.

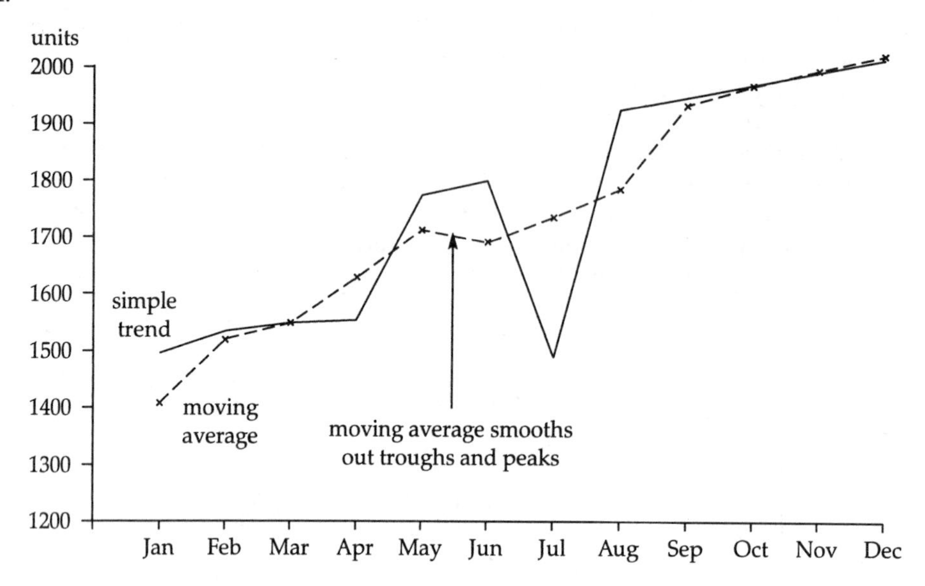

Figure A4 Simple trend and moving average graph

Index